Treatment of Perchlorate Using Single-Use Ion-Exchange Resins

Awwa Research Foundation

About the Awwa Research Foundation

The Awwa Research Foundation (AwwaRF) is a member-supported, international, nonprofit organization that sponsors research to enable water utilities, public health agencies, and other professionals to provide safe and affordable drinking water to consumers.

The Foundation's mission is to advance the science of water to improve the quality of life. To achieve this mission, the Foundation sponsors studies on all aspects of drinking water, including supply and resources, treatment, monitoring and analysis, distribution, management, and health effects. Funding for research is provided primarily by subscription payments from approximately 1,000 utilities, consulting firms, and manufacturers in North America and abroad. Additional funding comes from collaborative partnerships with other national and international organizations, allowing for resources to be leveraged, expertise to be shared, and broad-based knowledge to be developed and disseminated. Government funding serves as a third source of research dollars.

From its headquarters in Denver, Colorado, the Foundation's staff directs and supports the efforts of more than 800 volunteers who serve on the board of trustees and various committees. These volunteers represent many facets of the water industry, and contribute their expertise to select and monitor research studies that benefit the entire drinking water community.

The results of research are disseminated through a number of channels, including reports, the Web site, conferences, and periodicals.

For subscribers, the Foundation serves as a cooperative program in which water suppliers unite to pool their resources. By applying Foundation research findings, these water suppliers can save substantial costs and stay on the leading edge of drinking water science and technology. Since its inception, AwwaRF has supplied the water community with more than $300 million in applied research.

More information about the Foundation and how to become a subscriber is available on the Web at **www.awwarf.org**.

Treatment of Perchlorate Using Single-Use Ion-Exchange Resins

Prepared by:
Lee Aldridge, **Thomas Gillogly**, **Joan Oppenheimer**, **Geno Lehman**, **Kristie Witter**, and **Arturo Burbano**
MWH, 300 North Lake, Suite 1200, Pasadena, CA 91101

and

Dennis Clifford and **Anthony Tripp**
University of Houston, Department of Civil and Environmental Engineering
Houston, TX 77204-4791

Jointly sponsored by:
Awwa Research Foundation
6666 West Quincy Avenue, Denver, CO 80235-3098

U.S. Environmental Protection Agency
Washington D.C.

and

East Valley Water District
San Bernardino, CA

Published by:

Awwa Research Foundation
Advancing the Science of Water®

IWA Publishing

DISCLAIMER

This study was funded by the Awwa Research Foundation (AwwaRF). AwwaRF assumes no responsibility for the content of the research study reported in this publication or for the opinions or statements of fact expressed in the report. The mention of trade names for commercial products does not represent or imply the approval or endorsement of AwwaRF. This report is presented solely for informational purposes.

Published by IWA Publishing, Alliance House, 12 Caxton Street, London SW1H 0QS, UK

Telephone: +44 (0) 20 7654 5500; Fax: +44 (0) 20 7654 5555; Email: publications@iwap.co.uk;
Web: **www.iwapublishing.com**

AwwaRF report number 91038F. Originally published by AwwaRF for its subscribers in 2004.
IWA Publishing version published 2005.

ISBN 1 84339 931 8

CONTENTS

LIST OF TABLES

LIST OF FIGURES

FOREWORD

The Awwa Research Foundation is a nonprofit corporation that is dedicated to the implementation of a research effort to help utilities respond to regulatory requirements and traditional high-priority concerns of the industry. The research agenda is developed through a process of consultation with subscribers and drinking water professionals. Under the umbrella of a Strategic Research Plan, the Research Advisory Council prioritizes the suggested projects based upon current and future needs, applicability, past work; the recommendations are forwarded to the Board of Trustees for final selection. The foundation also sponsors research projects through the unsolicited proposal process; the Collaborative Research, Research Applications, and Tailored Collaboration programs; and various joint research efforts with organizations such as the U.S. Environmental Protection Agency, the U.S. Bureau of Reclamation, and the Association of California Water Agencies.

This publication is a result of one of these sponsored studies, and it is hoped that its findings will be applied in communities throughout the world. The following report serves not only as a means of communicating the results of the water industry's centralized research program but also as a tool to enlist the further support of the nonmember utilities and individuals.

Projects are managed closely from their inception to the final report by the foundation's staff and large cadre of volunteers who willingly contribute their time and expertise. The foundation serves a planning and management function and awards contracts to other institutions such as water utilities, universities, and engineering firms. Their funding for this research effort comes primarily from the Subscription Program, through which water utilities subscribe to the research program and make an annual payment proportionate to the volume of water they deliver and consultants and manufacturers subscribe based on their annual billings. The program offers a cost-effective and fair method for funding research in the public interests.

A broad spectrum of water supply issues is addressed by the foundation's research agenda; resources, treatment and operations, distribution and storage, water quality and analysis, toxicology, economics, and management. The ultimate purpose of the coordinated effort is to assist water suppliers in providing the highest possible quality of water economically and reliably. The true benefits are realized when the results are implemented at the utility level. The foundation's trustees are pleased to offer this publication as a contribution toward that end.

Walter J. Bishop
Chair, Board of Trustees
Awwa Research Foundation

James F. Manwaring, P.E.
Executive Director
Awwa Research Foundation

ACKNOWLEDGMENTS

The authors wish to thank Frank Blaha, project manager, of the Awwa Research Foundation (AwwaRF), and the following members of the Project Advisory Committee (PAC); Gerald Guter, BasinWater, Rancho Cucamonga, CA; and Jerry Lowry, Lowry Engineering, Blue Hill, Maine, for the opportunity to perform this work under the project continuation reserve program. This study expanded the findings of Project 2805, Treatability of Perchlorate in Groundwater Using Ion-Exchange Technology, to include the applicability of single-use ion-exchange resins to mitigate perchlorate contamination at municipal sites. It should be noted that the information presented herein is based on the views of the project team and not of Awwa Research Foundation.

The authors would also like to thank Calgon Carbon Corporation and US Filter for providing site lists and case studies of major perchlorate treatment installations utilizing ion-exchange technology. We are also indebted to the resin manufacturers for providing material cost information that was utilized in the engineering analysis.

EXECUTIVE SUMMARY

The continued exploration of ion-exchange perchlorate treatment has resulted in an increasing interest and implementation of single-use or "non-regenerable" ion-exchange resin systems. These systems generally employ resins with a high affinity for perchlorate in a simple fixed-bed configuration. Once exhausted, the resin is simply removed and replaced with fresh resin. The exhausted resin is then either regenerated at a centralized processing facility that is designed to manage the challenging regeneration procedure with potentially hazardous chemicals, or is destroyed in an incineration or co-generation facility. This report presents an evaluation of the:

- Relative impact of selected water quality parameters on effective life of resins. The volume of water or duration of time which a single-use resin system can effectively remove perchlorate is relatively less sensitive to the perchlorate concentration (1 to 200 μg/L) as compared to increasing concentration of chloride (5 to 200 mg/L), nitrate (0.5 to 20 mg/L as N), or sulfate (5 to 200 mg/L).

- Relative performance of resins in model waters. For all four model water qualities (see Table 3-5) evaluated, it was determined that the bifunctional resins consistently exhibited a longer effective life than the triethyl and tripropyl resins, which had longer lives than the Type I (trimethyl) resins. It was estimated that tripropyl resins had a 37 percent longer effective life than triethyl resins in low-nitrate or high-sulfate conditions, whereas they only lasted about 8 percent longer in high-nitrate or low-sulfate conditions. Furthermore, the triethyl resins lasted 3 to 9 times longer than the trimethyl resins for equivalent water qualities.

- Relative water production costs of single-use systems. The engineering analysis of each resin under the range of water qualities tested demonstrated that the resin with the longest effective life was not necessarily the most economical (see Figure 3-1). For three of the four model waters, the bifunctional resin had the highest water production costs. The triethyl and tripropyl resins generally had the lowest costs. However, for low-sulfate conditions, the Type I (trimethyl) resin was cost competitive.

- Relative costs of single-use to conventional ion-exchange systems. For a water quality typical of those encountered in southern California (ClO_4^- <20 μg/L, SO_4^{2-} 40 mg/L, NO_3^- 9 mg/L as N, Cl^- 30 mg/L, HCO_3^- 250 mg/L), it was estimated that single-use systems are typically more economical than regenerative systems (see Table 4-3).

- Full-scale implementation of single-use ion-exchange systems. Within California there are at least 17 full-scale installations utilizing single-use ion exchange systems with a total design capacity of nearly 60-MGD.

With the increasing demands on water supplies, the increasing detection of perchlorate-impaired waters, and the development of perchlorate regulations in several states, single-use ion-exchange resins have been shown to be more than just an effective interim measure. Depending on water quality, the costs of these single-uses can be substantially lower than conventional ion-exchange systems for perchlorate removal.

CHAPTER 1 – INTRODUCTION

The detection of perchlorate in California groundwaters in early 1997, prompted studies to evaluate treatment technologies. The technologies demonstrated to be technically feasible for drinking-water treatment applications are biological reduction, ion exchange, reverse osmosis (RO) membranes, and pretreated granular activated carbon (GAC). Of these technologies, ion exchange is the most proven and publicly accepted process capable of meeting existing treatment requirements. It is therefore the treatment process most frequently installed at sites with perchlorate contamination.

ION-EXCHANGE TREATMENT APPROACHES

A federal regulatory standard for perchlorate has not been established. However several states have set interim standards or action levels. Current perchlorate treatment installations typically optimize the design of perchlorate removal systems to allow production of finished water with perchlorate concentrations less than 4 μg/L. Because ion exchange is an established technology that has proven to be effective and reliable in removing perchlorate to below-detection-limits, it is frequently selected as a simple and cost-effective solution in the water treatment field. It can also provide simultaneous removal of other contaminants of concern, including nitrate, arsenic, uranium, or hexavalent chromium. Furthermore, many suitable commercial resins have already been certified for use in municipal applications. The three different ion-exchange approaches currently being utilized to treat perchlorate-contaminated water for potable use are briefly described below.

Regenerable Ion Exchange and Brine Disposal

Ion-exchange treatment using polyacrylic and polystyrene resins were found to be effective at removing perchlorate from contaminated groundwater (Clifford *et al.*, 2003). The primary concern surrounding the use of conventional ion exchange, however, is the specialized disposal and/or treatment of the perchlorate-laden saline waste stream generated as a by-product of the process. Current ion-exchange facilities typically use 6 to 12 percent NaCl brine solutions to regenerate resin exhausted with perchlorate. The spent brine is often disposed of in a dedicated non-reclaimable industrial waste-discharge line (sewer) established by local or regional waste treatment agencies specifically for highly saline wastes which will not be reclaimed for irrigation or other uses following waste treatment. Treatment costs associated with conventional ion-exchange treatment utilizing a brine disposal line can be costly due to large volumes of make-up water, salt required for regeneration, and the expense associated with the disposal of the brine. This approach, however, is also responsible for the reintroduction of the perchlorate-laden brine back into the environment if not removed by the waste treatment process. Generally speaking, waste treatment processes are not configured to remove perchlorate from the waste stream.

Regenerable Ion Exchange and Brine Treatment

An alternative means of handling the perchlorate-laden brine is implementing brine treatment and reuse strategies. Brine treatment could potentially eliminate the costly and specialized disposal and treatment of a concentrated waste stream, minimize the volume of waste discharged, reduce salt costs, minimize impact to the environment through perchlorate destruction, and prepare for possible future regulations that might require brine pretreatment prior to disposal. Three different brine treatment processes were investigated and pilot-tested as a part of AwwaRF Project 2805: biological treatment; a physical/chemical high temperature and pressure catalytic process; and electrolytic reductive process (Aldridge *et al.*, 2004). It was found that the biological brine treatment and the physical/chemical catalytic processes were effective strategies for treating and preparing perchlorate-laden brine for reuse as a regenerant solution. The electrolytic process evaluated, however, was unable to effectively remove perchlorate in spent brine.

Single-Use Ion-Exchange Resin Treatment Alternative

Recently, an increased interest in the use of "non-regenerable" or single-use resins for perchlorate removal has created an alternative solution for treatment of perchlorate-contaminated groundwater. Due to the high affinity for perchlorate, these single-use or disposable resins are not conventionally regenerable and are typically utilized in simple fixed-bed configuration. Once exhausted, the resin is simply removed and replaced with fresh resin. The exhausted resin is either regenerated at a centralized processing facility that is suited to managing the regeneration procedure and the handling of the associated regenerant chemicals or destroyed, typically at an incineration or co-generation facility.

PROJECT OBJECTIVES

The objective of this study was to update the findings of Project 2805, Treatability of Perchlorate in Groundwater Using Ion-Exchange Technology (Aldridge *et al.*, 2004), to include consideration of single-use ion-exchange resins for perchlorate removal.

This objective was achieved through the provision of the following sequence of tasks:

- Discussion of the types of single-use resins commercially available;
- Model predicted performance of these resins under varying water quality conditions;
- Engineering analysis of treatment costs for single-use resins;
- Comparison of costs against regenerable systems with and without brine treatment;
- Listing of perchlorate contaminated municipal sites installing sinlge-use systems;
- Use of a case study to compare relative costs of regenerable and non-regenerable systems.

CHAPTER 2 – SINGLE-USE RESIN APPROACH

Given the recent interest in single-use resins, this study investigates the feasibility of using resins with a high affinity for perchlorate (also called disposable) as an alternative ion-exchange process for perchlorate removal. This process takes advantage of the high perchlorate capacity of specifically selected resins and the reduction in costs that are a result of eliminating conventional regeneration. A review of commercially available resin types will be provided and evaluate their perchlorate removal capabilities. Data from both experimental and modeling studies will support this performance analysis. This section will provide the foundation supporting the basic engineering considerations used in process design and subsequent consideration of operation and maintenance conditions. Finally, methods and cost implications of resin disposal will be examined.

ION-EXCHANGE RESIN OVERVIEW

The strong affinity of specific ion-exchange resins for perchlorate is the basis of the engineering design for these applications. Resin selection plays a critical role since it will determine the economics and general feasibility of the process. A group of resins classified as Strong Base Anion (SBA) resins in the chloride form are most commonly considered for perchlorate removal. The major classification of SBA resins depends on the polymer matrix used. They include polyacrylic, polystyrene and polyvinylpyridene resins. The first two types are the most commonly used for drinking water applications.

Polyacrylic Resins

Figure 2-1 shows the resin matrix of a typical polyacrylic anion-exchange resin. The acrylic portion in the polymer chain results in an ion-exchange resin that is relatively hydrophilic. A trimethyl quaternary amine functional group is used and the cross-linking is accomplished with divinylbenzene. The selectivity of acrylic resins typically exhibits the following order: sulfate > perchlorate > nitrate > chloride> bicarbonate. During regular operation, perchlorate breakthrough will occur before sulfate but after nitrate. As shown in Figure 2-2, perchlorate breakthrough occurred after approximately 750 bed volumes (BV) of processed water. These results were obtained during pilot-scale evaluation of a polyacrylic resin for AwwaRF Project 2805. The perchlorate-contaminated groundwater contained 70 μg/L perchlorate, 6.3 mg/L nitrate as N, 30 mg/L sulfate, and 175 mg/L bicarbonate. Nitrate was observed in the effluent much earlier in the run at approximately 275 BV. Sulfate breakthrough was not observed during this period. The anticipated decrease in the effluent chloride ion concentration can be seen during the course of the exhaustion cycle as part of the multi-component equilibrium process and the breakthrough of bicarbonate, nitrate, and perchlorate ions.

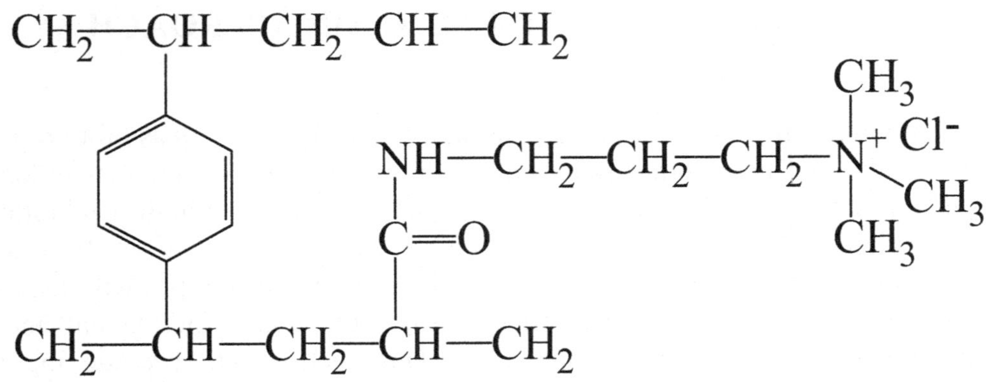

Figure 2-1 Resin matrix of a polyacrylic, trimethylamine, strong-base anion-exchange resin (Data from Harland 1994)

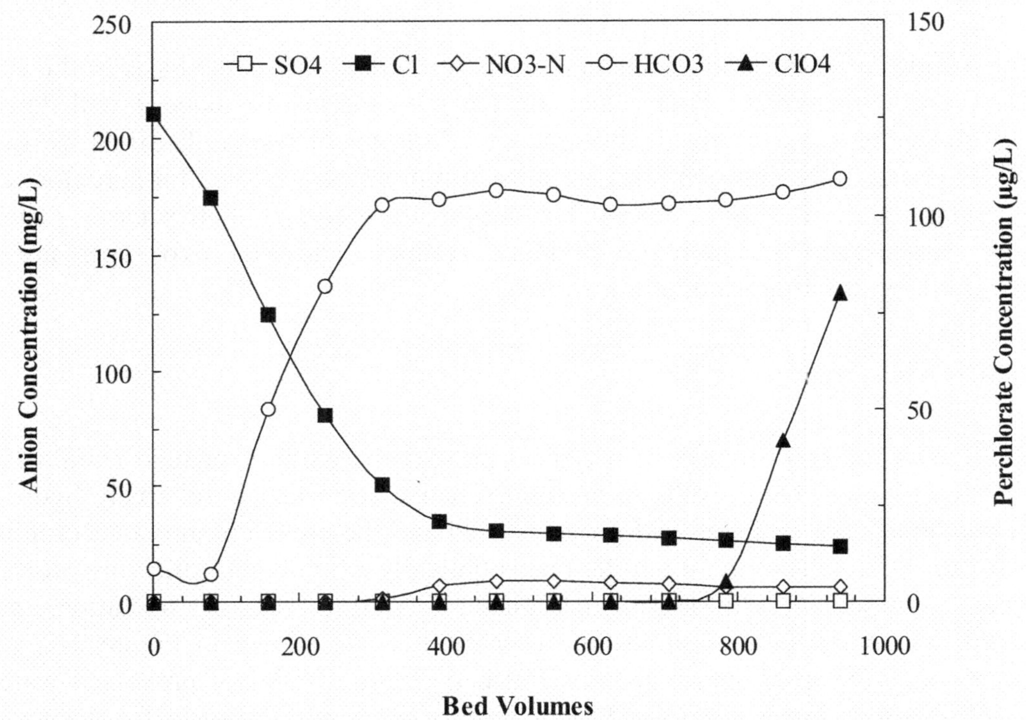

Figure 2-2 Exhaustion profile for evaluation of the polyacrylic resin (Source: Aldridge et al. 2004)

Counter-current regeneration of the polyacrylic resin was performed using a 6 percent sodium chloride solution until concentrations of perchlorate and nitrate remained below analytical detection limits. Figure 2-3 shows the profile of anions displaced during the regeneration cycle. The resin was regenerated with 16 BV of regenerant solution, equivalent to a salt loading rate of 62 lbs $NaCl/ft^3$ of resin. Nitrate, sulfate, and bicarbonate were completely removed from the resin after 6 BV of regeneration, and perchlorate was removed after 13 BV of regeneration.

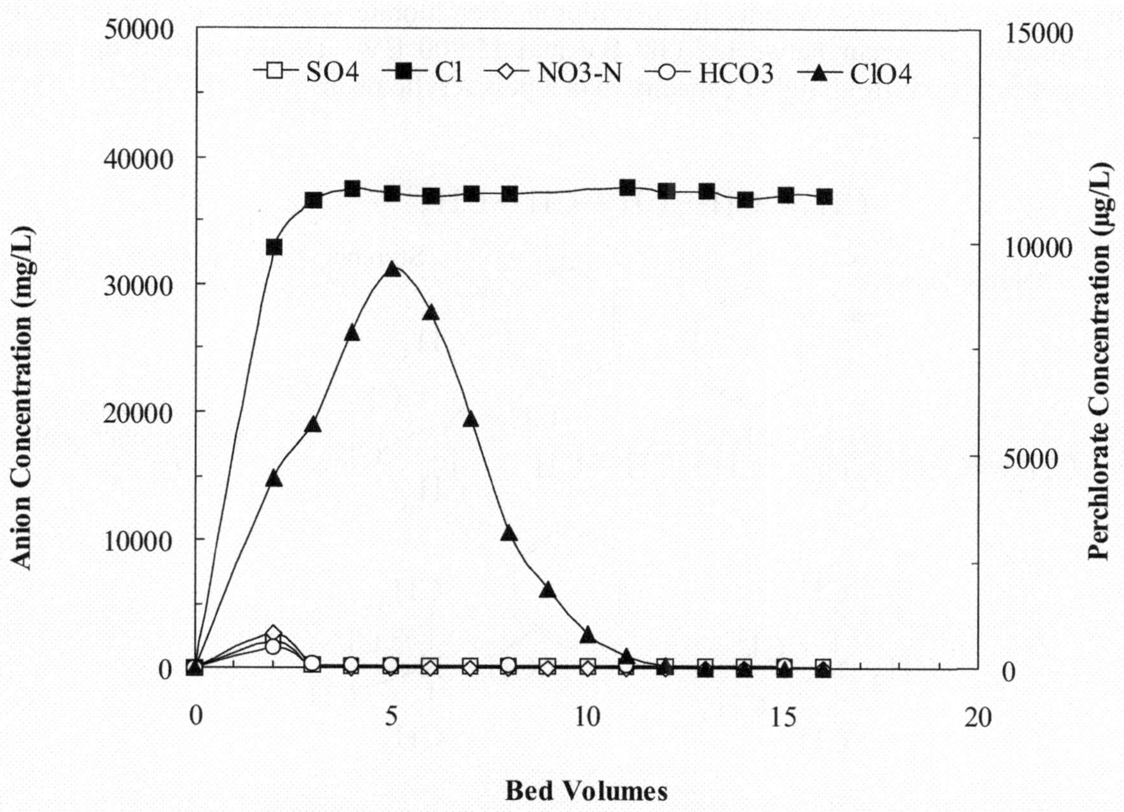

Figure 2-3 Polyacrylic resin regeneration evaluation - 62 lbs NaCl/ft^3 resin (Source: Aldridge et al. 2004)

These results demonstrated that polyacrylic resins are effective at removing perchlorate. However due to competition from other anions in solution, only approximately 800 BV of operation could be achieved before perchlorate breakthrough occurred. Additionally, it was found that the resin could be easily regenerated with a sodium chloride brine solution to remove all adsorbed perchlorate within 15 BVs. For these reasons, it is common for polyacrylic resins to be utilized in conventional ion-exchange plants due to their capacity for perchlorate and ease of regeneration.

Polystyrene Resins

Polystyrene resins exhibit a much higher affinity for perchlorate than for other anions. Styrenic resins can be further classified based on the functional groups included in their structure. Four categories are established: Type I, Type II, nitrate selective and bifunctional resins. Figure 2-4 is an example of an anion-exchange resin with a polystyrene resin matrix. This figure represents a standard Type I anion-exchange resin with a quaternary ammonium trimethylamine functional group attached to the aromatic ring of styrene. The polystyrene divinylbenzene matrix is considered to be nonpolar compared with matrices containing oxygen molecules (Tripp 2000).

Because of the stronger selectivity for perchlorate, perchlorate breakthrough for these styrenic resins typically can occur between 5,000 BV and 15,000 BV. During normal operation, this represents nearly a six-fold higher capacity than a polyacrylic resin.

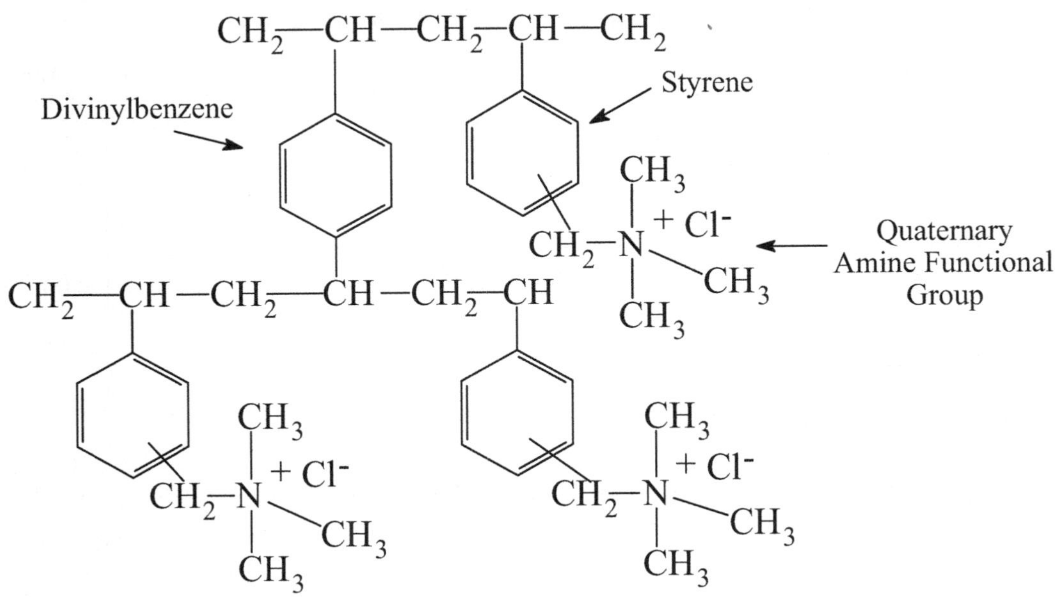

Figure 2-4 Resin matrix of a polystyrene, Type I, strong-base anion-exchange resin in the chloride form (Data from Helfferich 1962)

Type I / Type II

Type I resins are SBA resins including the basic trimethyl quaternary amine as their quaternary amine functional group, while Type II resins include dimethylethanolamine instead (Tripp 2000). Figure 2-5 represents the two functional groups used in the Type I and Type II resins. As a result of the inclusion of an ethanol group for one methyl group, Type II resins are slightly more hydrophilic than Type I resins. Both types have greater affinity for sulfate relative to nitrate.

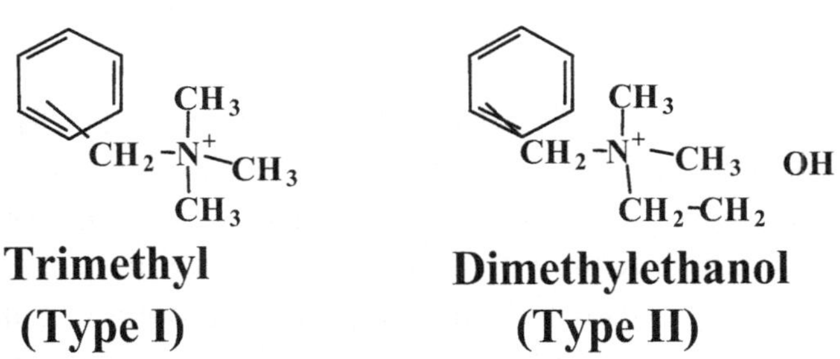

Figure 2-5 Quaternary amine functional groups for Type I and Type II polystyrene resins (Data from Dorfner 1991)

A Type I polystyrene resin was operated in the field with perchlorate-contaminated groundwater for a total of 7600 BV, during which time, the breakthrough of perchlorate, nitrate, sulfate and bicarbonate were all observed (AwwaRF Project 2805). Perchlorate was observed in the effluent after approximately 5700 BV, as shown in Figure 2-6. Bicarbonate, nitrate and sulfate were also detected in the effluent at approximately 100 BV, 400 BV, and 1500 BV, respectively.

Counter-current regeneration of the polystyrene resin was performed using a 6 percent NaCl solution. Figure 2-7 shows the regeneration profile of anions adsorbed during the exhaustion cycle. Nitrate, sulfate, and bicarbonate were completely removed from the resin after 8 BV of regeneration. Figure 2-8 shows the regeneration profile of perchlorate. The perchlorate concentration in the regenerant brine peaked at approximately 6000 µg/L after 2 BV of regeneration. However, even after 140 BV (530 lbs NaCl/ft^3 resin) of regeneration, analysis of the brine indicated that approximately 800 µg/L perchlorate was still being displaced from the styrenic resin. Regeneration of the polystyrene resin was terminated after 140 BV because the profile demonstrated the large quantity of salt required for the complete resin regeneration.

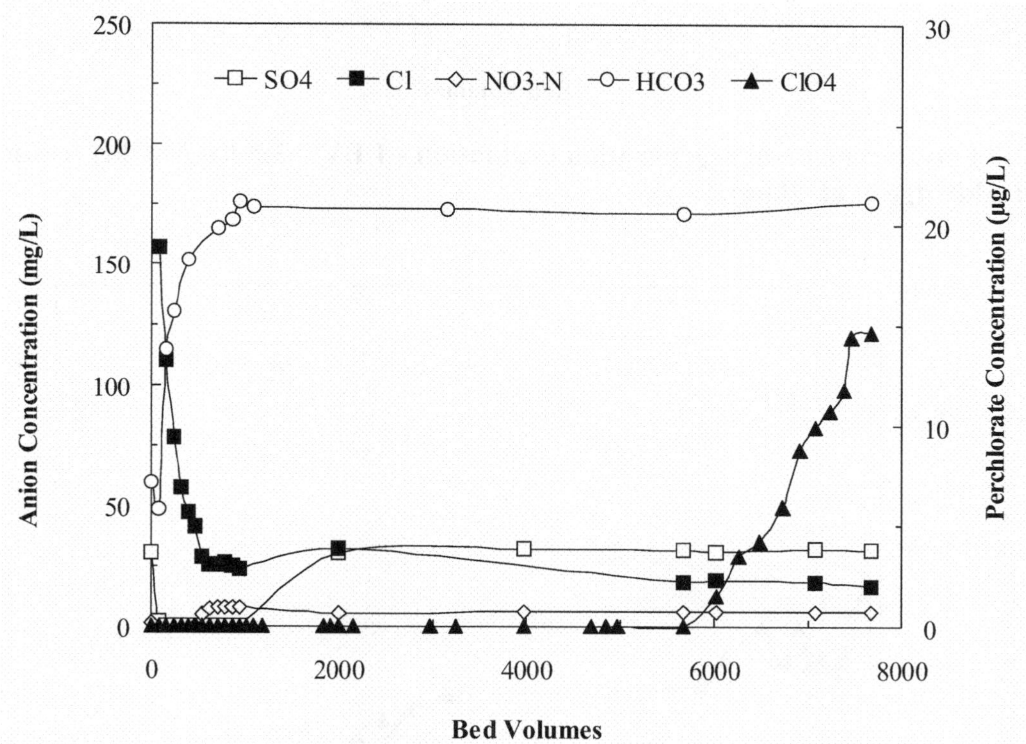

Figure 2-6 Exhaustion profile for the initial evaluation of the polystyrene resin (Source: Aldridge et al. 2004)

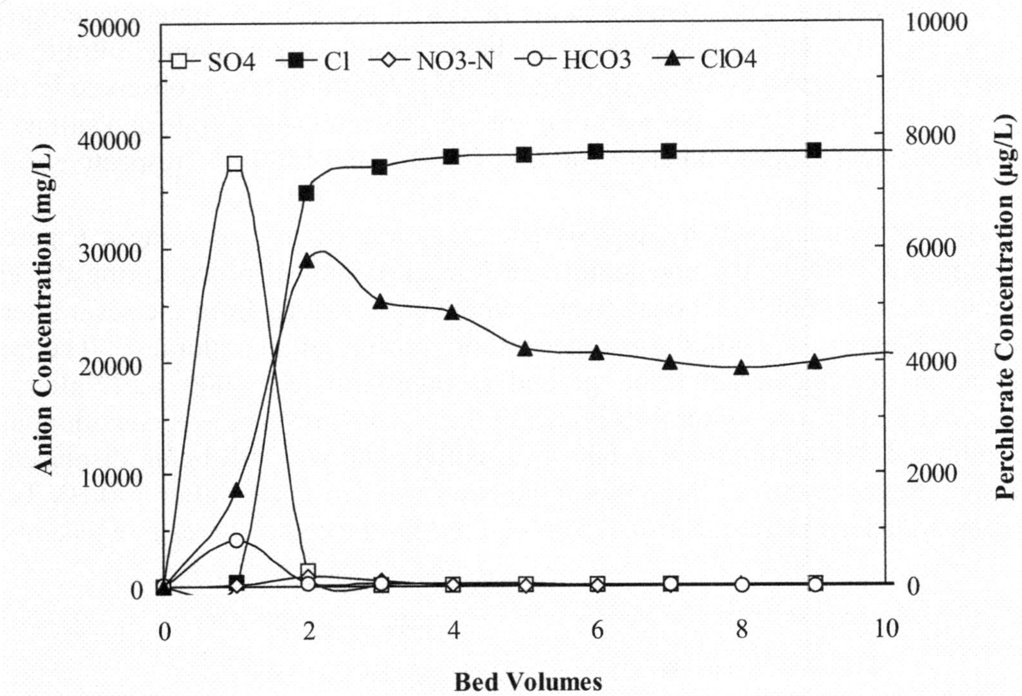

Figure 2-7 Polystyrene resin regeneration evaluation - 1 BV = 3.8 lbs NaCl/ft^3 resin (Source: Aldridge et al. 2004)

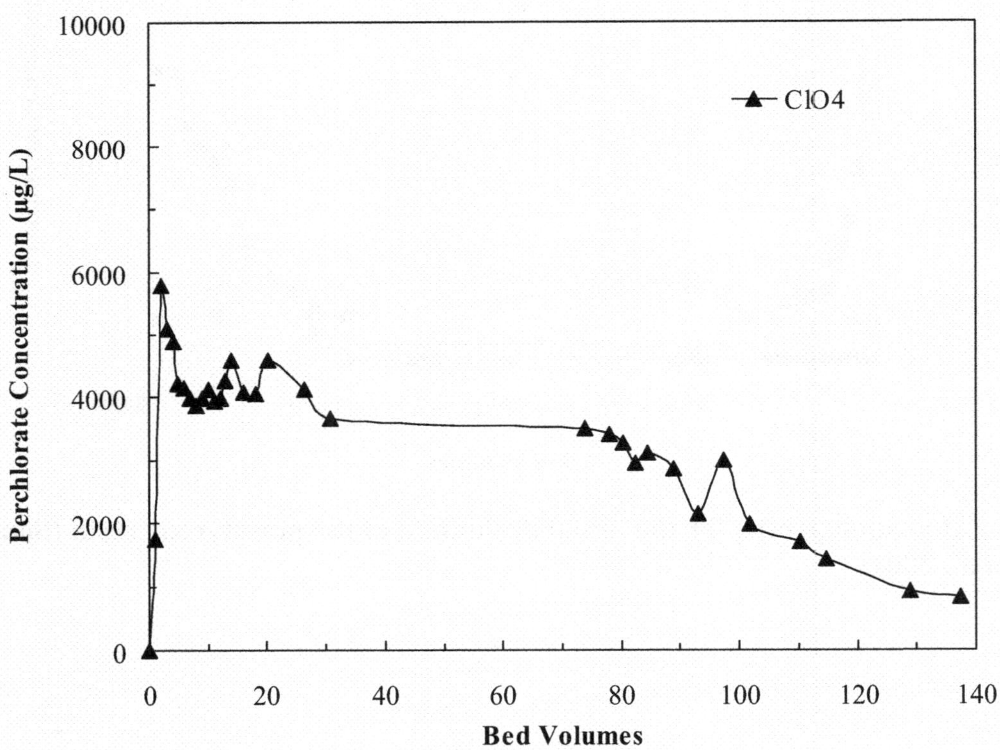

Figure 2-8 Perchlorate elution during polystyrene resin regeneration – 140 BV/530 lbs NaCl/ft^3 resin (Source: Aldridge et al. 2004)

These results also demonstrated that although the polystyrene resin could be operated for extended run lengths, complete regeneration using a conventional sodium chloride brine solution required large quantities of salt. Even after a salt loading rate of 530 lbs/ft^3, perchlorate was still being removed from the resin. The polystyrene resin has a perchlorate run length nearly eight times longer than the polyacrylic resin, but regeneration requires almost ten times the salt. Consequently, the use of a polystyrene resin may not be feasible or economical given that extremely large quantities of salt would be required for complete regeneration. However, application of these polystyrene resins, with their high perchlorate selectivities, may be more suitable for a single-use approach, in which the exhausted resin is disposed of rather than regenerated.

Nitrate-Selective Resins

Nitrate over sulfate selective (NSS) resins have specific functional groups, including triethylamine and tripropylamine (Figure 2-9), that provide the resin with a higher affinity for nitrate than sulfate. The increase in length of the alkyl group causes a decrease in the selectivity of the resin for divalent ions and an increase in selectivity for monovalent ions. Resins with triethyl and longer functional groups tend to be less selective for sulfate and more selective for nitrate (Tripp, 2000).

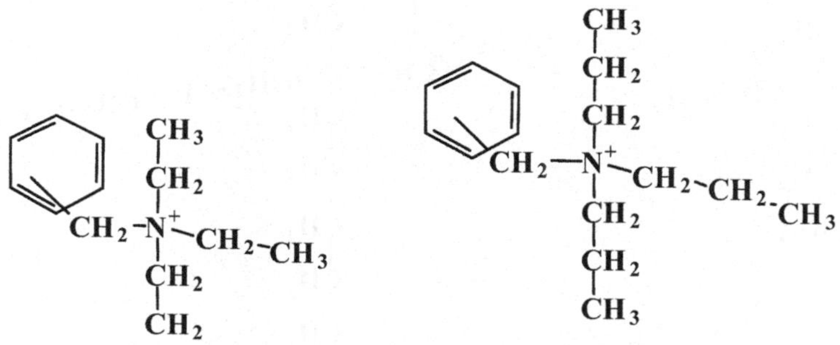

Figure 2-9 Quaternary amine functional groups for nitrate-selective polystyrene resins (Data from Dorfner 1991; Li and SenGupta 2000)

In a previous study (Tripp 2000) the time to perchlorate breakthrough was determined for a commercially available NSS resin. For this synthetic groundwater (500 µg/L ClO_4^-, 30 mg/L Cl^-, 6 mg/L NO_3^- as N, 122 mg/L HCO_3^-) with a moderate concentration (50 mg/L) of sulfate, perchlorate was detected after 10,000 BV. Continued operation of the NSS resin demonstrated that the resin could operate for approximately 60,000 BV until complete perchlorate exhaustion.

Bifunctional resins

Specially designed resins have been experimentally developed for removing large, poorly hydrated ions, like perchlorate. These resins show much higher selectivity for perchlorate than for any of the major competing anions. Perchlorate capacities as high as 110,000 BV have been reported.

Certain types of these anion-exchange resins are usually called "bifunctional resins" because they include two different quaternary amine functional groups in their structure. Typically, trihexylamine and either tripropyl-, or triethyl-amine functional groups are used (Figure 2-9). One of them is responsible for the high perchlorate selectivity, which is linked to the presence of a long alkylic chain (trihexylamine functional group). Conversely, the other group has a short alkylic chain (i.e., tripropylamine or triethylamine), that improves the kinetics of the ion-exchange process. Previous studies have shown that compared to Type I SBA monofunctional resins, bifunctional resins have been shown to have perchlorate capacities at least five-fold higher. This difference is impacted by several factors, including: water chemistry, sizes of functional groups of the resin, and initial perchlorate concentration.

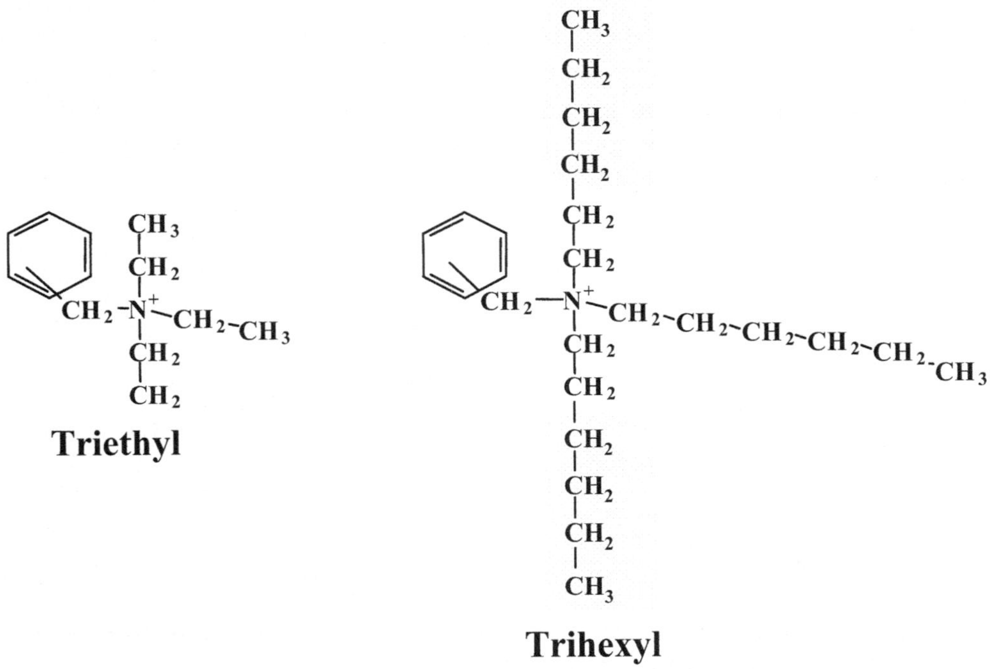

Figure 2-10 Quaternary amine functional groups for bifunctional resins (Data from Dofner 1991)

MATHEMATICAL MODELING

Studies have shown that the resin's affinity/selectivity for the perchlorate ion versus other competing ions such as chloride, sulfate or nitrate is key to determining the efficiency of the overall process (Subramonian 1988). It is important to emphasize that perchlorate is only a trace compound found in concentrations typically several orders of magnitude lower than the other competing anions. Consequently, a high percentage of the active sites on the resin structure will be occupied by competing anions and small variations in the concentration of such anions may have large impacts on the resin's perchlorate removal capacity. Site-specific water quality knowledge is a critical parameter in choosing the appropriate ion-exchange resin.

In terms of water quality, the concentration of sulfate, nitrate, perchlorate, and chloride will have the most significant impact on the number of bed volumes to perchlorate breakthrough with bicarbonate having a lesser effect. For a specific resin, capacity and perchlorate selectivity will have the greatest impact on perchlorate breakthrough, with sulfate and nitrate selectivity having secondary impact. A useful tool to help estimate the effective number of bed volumes to perchlorate breakthrough is the application of mathematical models based on multi-component equilibrium theory. Two commonly used ion-exchange mathematical models were considered: the EMCT program developed at the University of Houston and the commercially available IX WINDOWS PRO program.

Modeling Assumptions

The mathematical models document the effect that variations in the concentration of sulfate, nitrate, chloride, and perchlorate have upon the bed volumes to perchlorate breakthrough of Type I (trimethyl), triethylamine (TEA), tripropylamine (TPA), and bifunctional (trihexyl-triethyl amine mix) polystyrene ion-exchange resins.

No specific brand of resin was used for this modeling effort, but knowledge of the specific chemical composition of these resins was applied to the model. The behavior of these resins toward perchlorate was established during (Phase I of 2805). As the alkyl chain of the functional group increases (e.g. methyl, ethyl, propyl) the selectivity of a resin toward perchlorate and nitrate increases while the selectivity towards sulfate decreases. All of the four resin types used for modeling have perchlorate as the most selective anion being tested. Other trace anions, such as chromate (CrO_4^{2-}) and uranyl carbonate ($UO_2(CO_3)_3^{4-}$), may have higher selectivity than perchlorate, but they are not considered in the models. Nitrate is less preferred than sulfate for the Type I resin, but more preferred in the other three resins. Modeled resin characteristics in terms of perchlorate selectivity (relative to chloride at 20°C), functional group, and the total capacity of the resins are listed in Table 2-1.

Table 2-1
Characteristics of the resins used in the model

Functional Group	Perchlorate Selectivity	Nitrate Selectivity	Sulfate Selectivity	Resin Capacity (meq/L)
Trimethyl (Type I)	100	3.5	12	1.3
Triethyl	600	10	5	0.9
Tripropyl	800	12	4	0.8
Bifunctional	1000	15	3.5	0.9

The duration of a model run or bed volumes to perchlorate breakthrough will vary with the chemical composition of the source water. An increase in the amount of an anion in solution will shorten the run length, given that the amount of the other anions remains constant. The amount of the decrease will depend on the specific anion being increased. For modeling purposes the concentration of four components were varied to determine their specific effect on perchlorate run length. The base composition of the defined source water was considered to be constant with the exception of the specific anion being varied. The composition of the defined source water is given in Table 2-2 along with the composition range of the four specific anions varied for modeling purposes. The composition of bicarbonate was not varied since large changes in concentration will result in only small changes in perchlorate run length due to the low selectivity of the resins for bicarbonate.

Table 2-2
Composition of defined source water

Anion	Concentration meq/L (mg/L)	Range in Models meq/L (mg/L)
Perchlorate	0.0002 (0.02)	0.00001 – 0.002 (0.001-0.2)
Chloride	0.6 (21)	0.141-5.642 (5 – 200)
Nitrate	0.3 (4 as N)	0.036-1.429 (0.5-20 as N)
Sulfate	1 (48)	0.1 – 4 (4.8 – 192)
Bicarbonate	2 (122)	Not varied

The EMCT model was used to predict perchlorate run lengths assuming ideal operation of the columns. The model assumes there will be no mass transfer limitations, resulting in a square wave concentration versus BV output. However, in reality the columns are not ideal and there are mass transfer limitations that lead to early leakage of perchlorate, resulting in sinusoidal concentration versus bed volume output. Another computer program, the IX WINDOW PRO model, does have the ability to simulate mass transfer limitations to estimate leakage during the exhaustion of the resin. However, the user must choose the number of equilibrium stages to estimate leakage. The fewer equilibrium stages estimated will result in a greater apparent leakage. For these modeling purposes, ten-stages were chosen to estimate leakage. It is important to note that both EMCT and IX WINDOW PRO are equilibrium models. The EMCT program reports the theoretical run duration for square-wave front isotherm models. This value corresponds to the inflection point in the effluent curve of the IX WINDOWS PRO program.

There will be leakage during the exhaustion process. This leakage will have an effect on how the treatment facility is designed and the number of columns needed to guarantee that perchlorate is removed to below the MCL. This leakage does not significantly affect the calculation of run length to complete exhaustion when the system as a whole is considered over a significant period of time. When considered over time the IX WINDOWS PRO and the EMCT program yield similar run lengths to complete exhaustion. The reason the EMCT program is used is because it instantly gives the results of complete exhaustion. The EMCT program greatly reduces the amount of time required to provide similar results as the IX WINDOW PRO program. For comparison, the IX WINDOWS PRO program was used to model the exhaustion of a bifunctional resin type with the defined water composition. The profile of the concentration of perchlorate leaving the column is shown in Figure 2-11. The square-wave EMCT predicted run length to perchlorate breakthrough is also shown as the vertical line at 102,000 bed volumes. This illustrates the general agreement between the two programs.

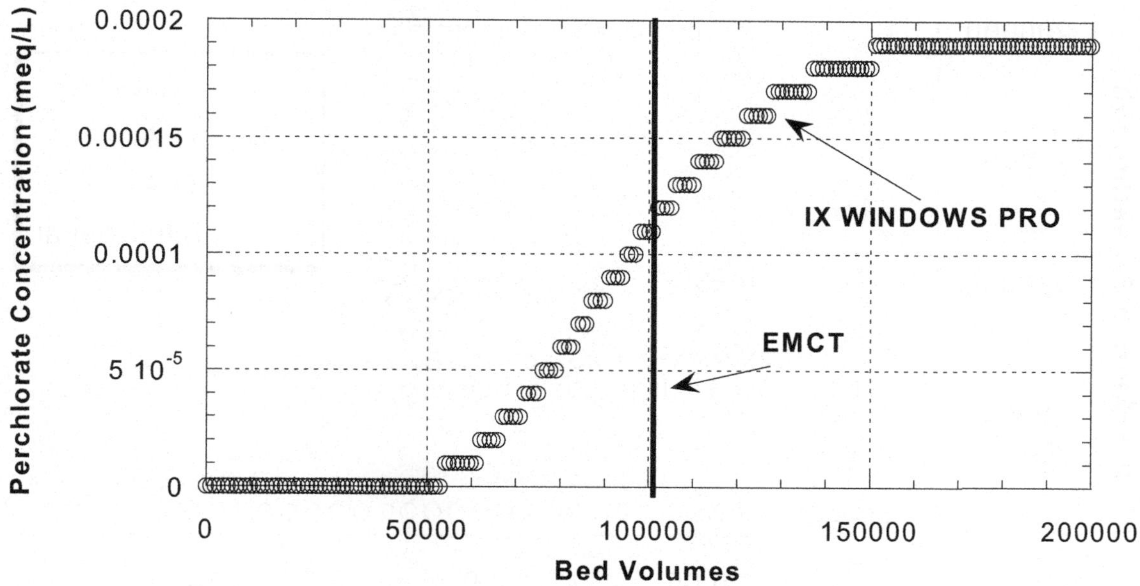

†To estimate leakage the number of plates used in the IX PRO models was ten (10).

Figure 2-11 Comparison of computer predicted perchlorate effluent curve for IX WINDOWS PRO and EMCT programs using basic defined source water composition

Modeling Results

The results of the modeling are presented graphically in Figure 2-12 through Figure 2-19. The even numbered figures show the change in the number of bed volumes to perchlorate breakthrough. The odd numbered figures depict the same information normalized to the highest number of bed volumes to perchlorate breakthrough for a specific resin type.

Modeling Discussion

The effect that compositional change in any single water quality component will have on bed volumes to perchlorate breakthrough is, in general, dependent upon the selectivity of the resin for that component. For example, if a resin has a greater selectivity for perchlorate than another resin, then changes in perchlorate concentration will have a greater effect on that resin. This effect is shown in Figure 2-12 and Figure 2-13. Figure 2-14 and Figure 2-15 illustrate this effect for chloride, Figure 2-16 and Figure 2-17 for nitrate, and Figure 2-18 and Figure 2-19 for sulfate.

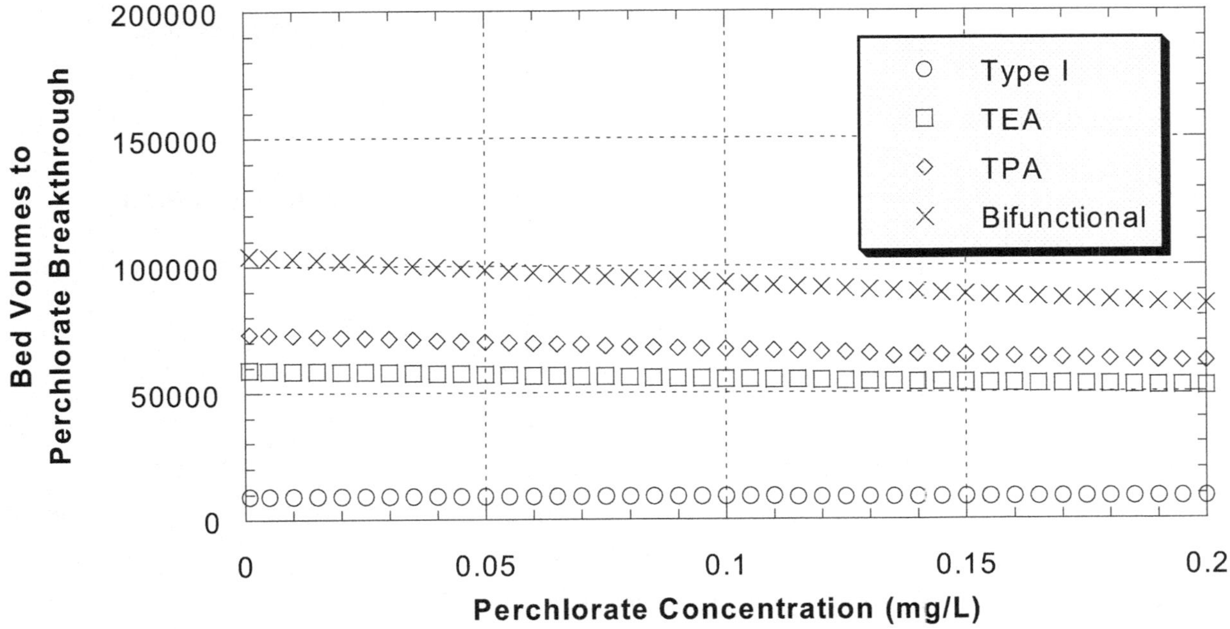

Figure 2-12 Effect of changing perchlorate concentration on run length to perchlorate breakthrough for four different types of polystyrene ion-exchange resins

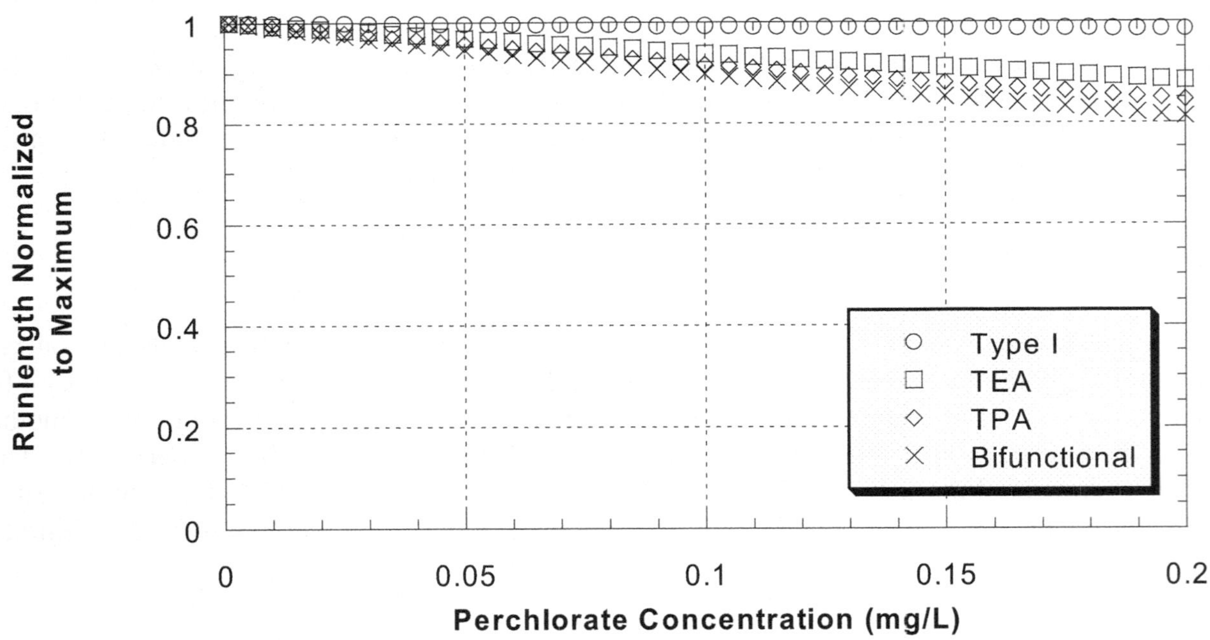

Figure 2-13 Effect of changing perchlorate concentration on run length to perchlorate breakthrough normalized to the maximum run length for each resin type

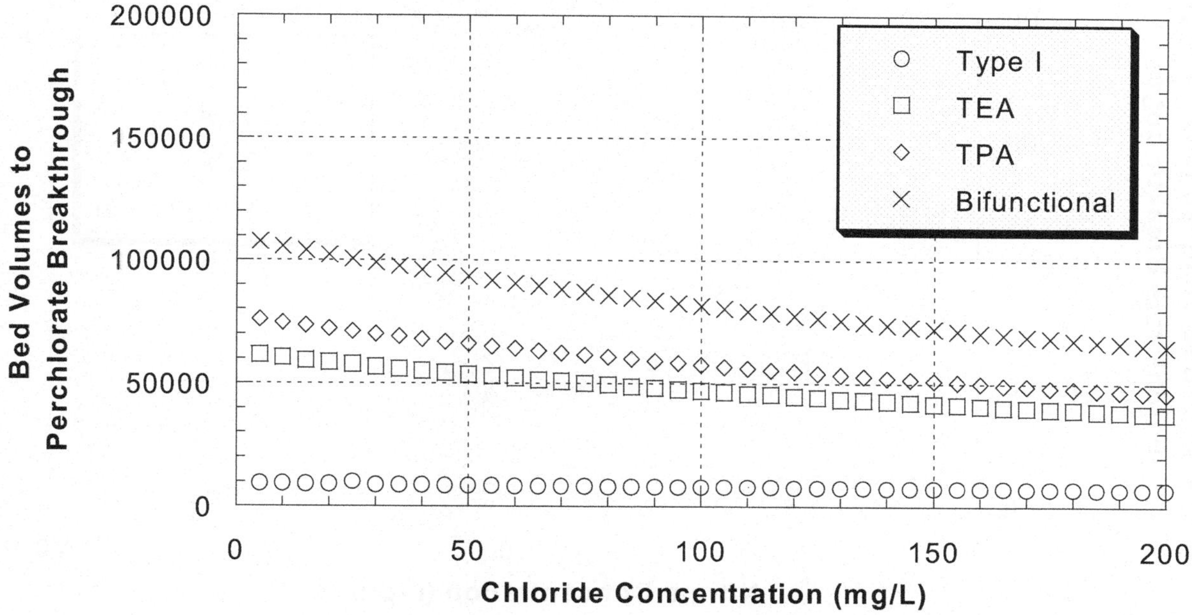

Figure 2-14 Effect of changing chloride concentration on run length to perchlorate breakthrough for four different types of polystyrene ion-exchange resins

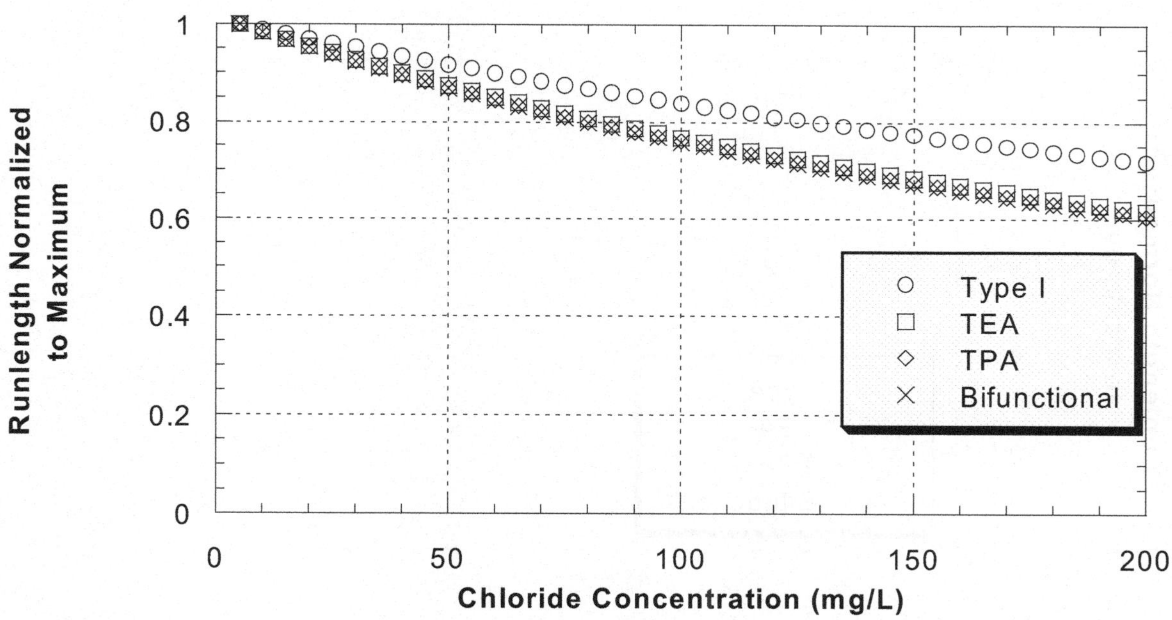

Figure 2-15 Effect of changing chloride concentration on run length to perchlorate breakthrough normalized to the maximum run length for each resin type

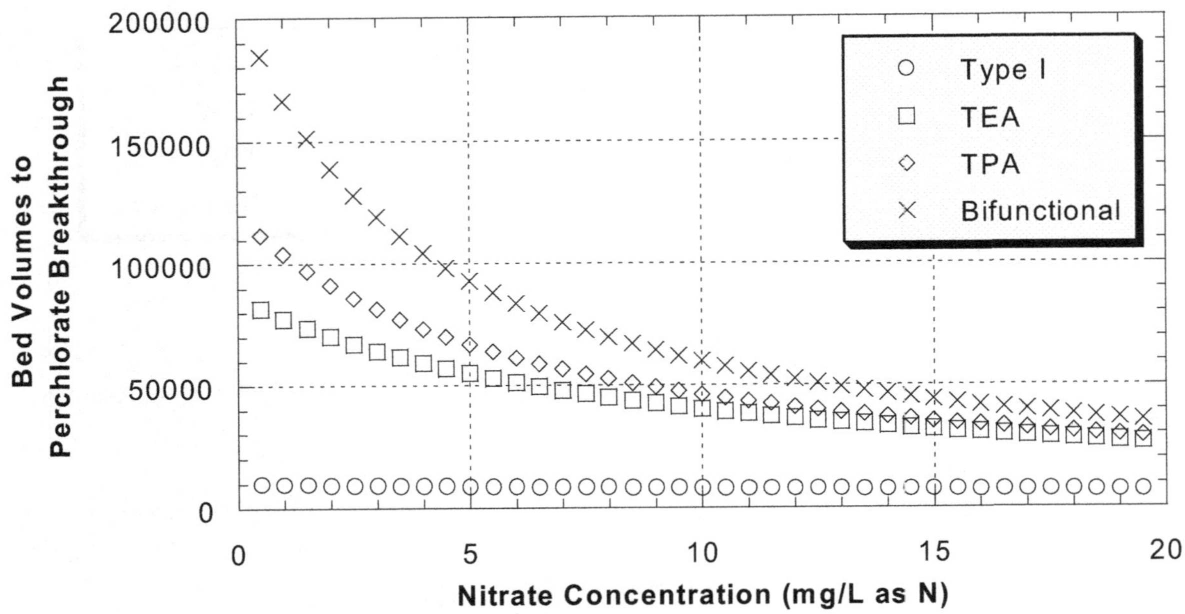

Figure 2-16 Effect of changing nitrate concentration on run length to perchlorate breakthrough for four different types of polystyrene ion-exchange resins

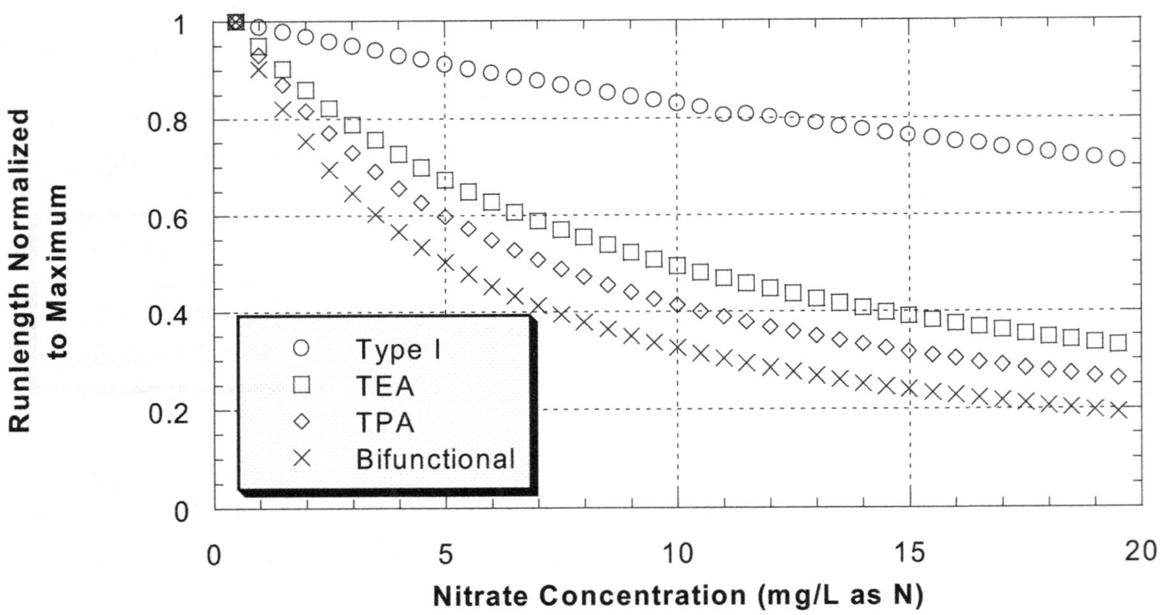

Figure 2-17 Effect of changing nitrate concentration on run length to perchlorate breakthrough normalized to the maximum run length for each resin type

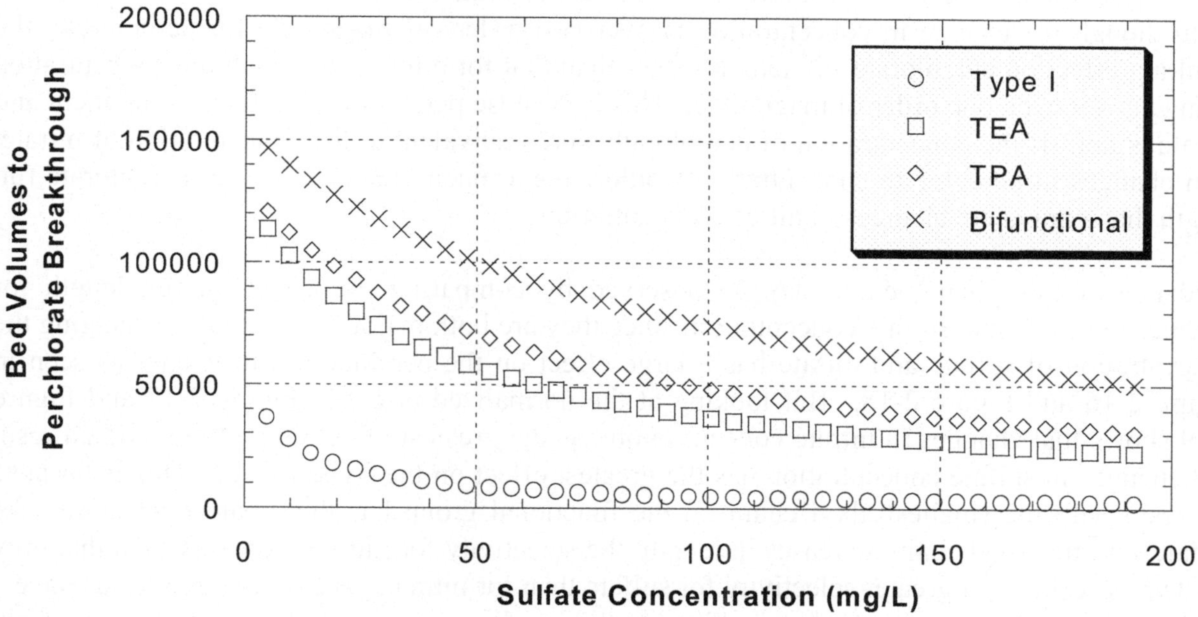

Figure 2-18 Effect of changing sulfate concentration on run length to perchlorate breakthrough for four different types of polystyrene ion-exchange resins

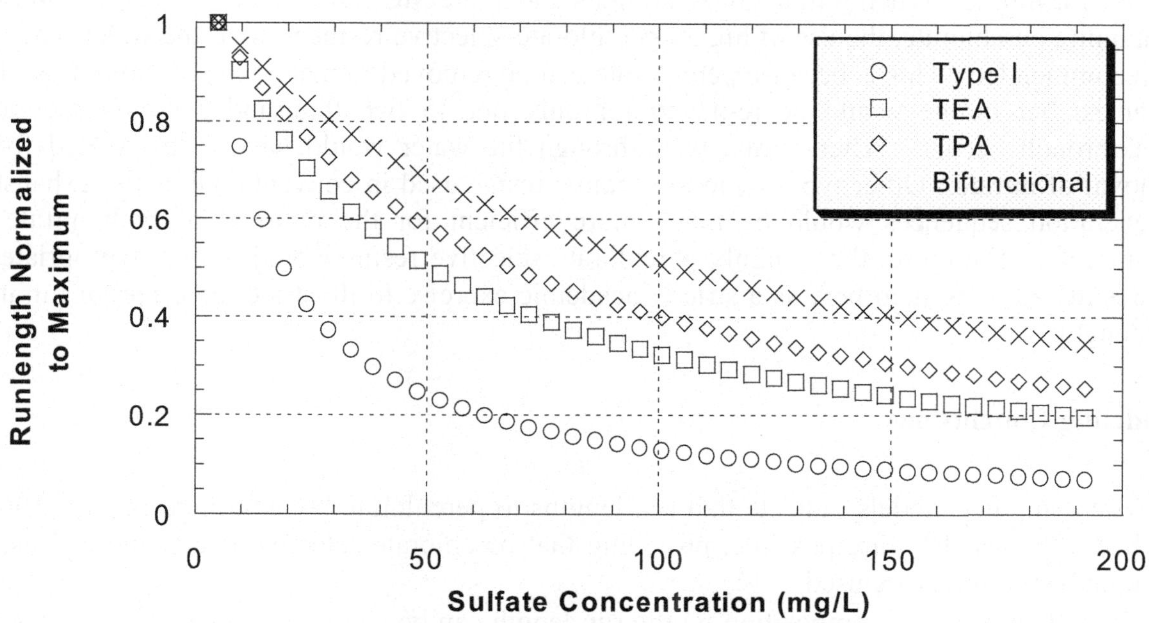

Figure 2-19 Effect of changing sulfate concentration on run length to perchlorate breakthrough normalized to the maximum run length for each resin type

The magnitude of change is also a function of the concentration of the various anions in solution. Even though the change in concentration is over two orders of magnitude for perchlorate, the resulting effect on perchlorate run length is less than that for nitrate, for which the concentration change is not even one order of magnitude. This is because perchlorate is a trace component and nitrate is not. If the concentration of perchlorate were equivalent to the concentration of nitrate, then changes in perchlorate concentration would have a much greater effect on perchlorate run length than equivalent changes in nitrate concentration.

Evidence of this phenomenon may be observed by comparing the effect on run length for changes in nitrate and sulfate concentration since they are both major components. Changing the concentration of sulfate and nitrate has a large effect on the perchlorate run length, as seen in Figure 2-16 and Figure 2-18. But looking at the normalized effect in Figure 2-17 and Figure 2-19 shows that changes in nitrate concentration has the greatest effect on the bifunctional resin and changes in sulfate concentration has the greatest effect on the Type I resin. This is because the resin with the smallest alkyl chain on the functional group has the greatest selectivity for sulfate. As the alkyl chain increases in length, the selectivity for nitrate increases such that only the Type I resin has a greater selectivity for sulfate than for nitrate. The other three resins have a greater selectivity for nitrate than sulfate, with the disparity increasing as the alkyl chain increases in length.

The effect of nitrate concentration upon perchlorate run length is illustrated in Figure 2-16 and Figure 2-17. In these figures the concentration of nitrate was varied from 0.5 to 20 mg/L as nitrogen. The upper value of this range is twice the 10 mg-N/L maximum contamination level (MCL) for nitrate. When nitrate concentrations above the MCL are encountered in groundwater containing perchlorate, the use of highly perchlorate-selective resins to treat the water would not be recommended. This is because perchlorate can be removed to many tens of thousands of bed volumes, but nitrate would be removed for only one to two thousand bed volumes before breakthrough occurs. After nitrate breakthrough the water would have to be treated for the removal of nitrate. Other types of ion-exchange resins used in conventional cyclic exhaustion-regeneration sequences would be much more efficient for the removal of both nitrate and perchlorate. Therefore, these highly perchlorate selective resins would not be appropriate and the results are presented here as a strictly academic exercise to illustrate the effect of nitrate on run length.

Modeling Conclusions

- The computer models indicate that run lengths to perchlorate breakthrough of over 100,000 bed volumes (BV) are possible, providing that perchlorate-selective resins and at least two columns in series are used.
- The effect of water composition on the run length can be significant for variations in sulfate and nitrate, with less impact from variations in perchlorate and chloride concentration change.
- Resins with the highest perchlorate selectivity are impacted most by a change in the water composition. These resins, however, experienced the longest run lengths to perchlorate breakthrough for all test conditions.

DESIGN

Currently installed full-scale single-use ion-exchange systems have shown promising results for perchlorate removal from wells in California and Nevada. The main design considerations for these systems are focused on the impact of water chemistry on resin capacity and leakage, economics of the process, and the particular requirements of a given application (i.e., flow rate, reliability, permitting). The considerations on the first topic (impact of water quality) were discussed in the previous section.

Regulatory approval for the installation of single-use resins typically requires "lead-lag" operation mode. An example of ion-exchange vessels in a lead-lag configuration is shown in Figure 2-20. The treatment equipment typically consists of two pressure vessels (#1 and #2) operated in series. Once the resin in the vessel #1 is saturated with perchlorate, the treatment unit is taken offline. After the spent resin is removed and replaced with fresh resin, vessel #2 takes its place as the lead vessel in the treatment system, vessel #1 becomes the lag vessel, and the treatment cycle continues. When using this configuration to remove perchlorate, bifunctional resins with equivalent capacities as high as 110,000 BV have been reportedly achieved.

Figure 2-20 High pressure vessels configured in a lead-lag configuration (courtesy of USFilter)

RESIN DISPOSAL

Once the resin is fully exhausted, it is typically removed from the pressure vessel by mechanical loading using a slurry truck (Figure 2-21). The spent resin is then transported to a handling facility where it is de-watered and transferred into a disposal container (typically 1000 L SuperSacks). The bulk stored spent resin is then hauled to an appropriate co-generation waste-to-energy or incineration facility, where it is destroyed.

It is less convenient from all points of view to fill landfills with materials that could be readily burned (i.e., oxidized). As in all waste burning, there are inorganic constituents in the off-gas (mainly chlorine, sulfur, and nitrogen). Co-generation and incineration facilities already have equipment in place to handle these inorganic compounds (i.e., as acid gases or nitrous oxides) and, therefore, require no additional pollution control equipment is needed to reduce or eliminate the contaminants of concern.

Perchlorate-laden spent resin is typically considered non-hazardous, non-DOT waste. There are no additional costs associated with handling or transporting and it can be destroyed in co-generation waste-to-energy plants that use "clean" waste to generate heat. The feed to the burner is not metered because all the generated heat is used to produce steam. The high-energy capacity of the spent ion-exchange resin (~ 14,000 Btu/lb) facilitates combustion and provides the needed BTUs to burn additional waste. Considering the technology trends, co-generation systems will remain as part of the national "waste disposal" options, given the detrimental consequences of landfill disposal.

Figure 2-21 Slurry truck for transportation of perchlorate-laden resin (courtesy of USFilter)

If spent ion-exchange resin is classified as a hazardous waste (due to accumulation of site-specific constituents such as arsenic, uranium, hexavalent chromium, etc. on the resin), high temperature incineration may be required. Incineration requires a carefully metered feed of materials based upon their energy capacity, in order to maintain the extremely high temperatures required for achieving complete or near-to-complete combustion. Secondary fuel(s) and additional incineration stages are required to provide the destruction of the trace organics that may be generated from the primary incineration. Since this feed requires significant analysis, materials can only be fed to the burner at a very slow rate, which makes the costs for incineration extremely high. High temperature incineration may not be permitted everywhere, mainly because the types of recalcitrant materials that are treated in typical incineration facilities generally cause deep concerns from local landholders

OPERATION AND MAINTENANCE CONSIDERATIONS

Mathematical modeling demonstrated that over 100,000 bed volumes to perchlorate breakthrough are possible when perchlorate-selective resins are used. Depending on the site-specific water quality constituents, the operation run time of a resin bed could easily extend from several weeks to months. With this in mind, the effect of minor components such as dissolved organic carbon (also removed by ion-exchange resins), particulates, microorganisms, and highly preferred inorganic species, such as chromate and uranium, could be significant and must be considered in the design of any treatment facility. Installation of biological control devices before and after the resin vessels is highly recommended to prevent potential biological growth, which will have a detrimental impact on the useful life of the media.

The impact of dissolved uranium could be important in the disposal of these resins when exhausted. These SBA resins generally have a high selectivity for uranium. Trace amounts of uranium concentrated by 100,000 fold could represent disposal problems. Other trace components such as chromium, arsenic, and selenium could also be a problem although their selectivity by the resins is much less than that for uranium or perchlorate and their resulting concentration factor would therefore be comparatively lower.

SUMMARY

Taking into account the information presented in this chapter (including the output from the mathematical model, O&M considerations, resin destruction, and other design considerations), an engineering analysis was performed to estimate the water production cost using a single-use resin approach. These results are presented in the following chapter and include water production costs associated with the application of Type I, tripropyl, triethyl, and bifunctional polystyrene resins for a variety of water quality conditions. The engineering analysis also includes a comparison of water production cost for single-use resin versus conventional ion exchange with and without brine treatment to determine the most cost-effective method of perchlorate removal.

CHAPTER 3 – ENGINEERING ANALYSIS

Given that only a limited number of full-scale single-use resin systems are currently in operation, an engineering analysis was performed to develop preliminary cost estimates of single-use systems for a range of water quality conditions. Additionally, treatment costs of a single-use system for a defined water quality were compared to the costs of a conventional regenerable ion-exchange system operated with brine disposal or brine treatment alternatives.

DESIGN AND COST INFORMATION

Site-specific data is essential for an engineering analysis, including design criteria, capital costs, annual O&M cost, and construction related costs. If this data is not available, assumptions may be made. Since the engineering analysis presented in this chapter is based on a theoretical system, all assumptions used for the analysis have been defined.

Design Conditions

The design conditions included in the engineering analysis are the following:

- Design flow rate
- Design run time (estimated number of bed volumes)
- Vessel diameter
- Service loading rate (gpm/ft^2)
- Headloss rate
- Maximum allowable headloss

These variables were used to calculate the number of IX vessels required under the specified design conditions.

The diameter of IX vessels was assumed to be 10 ft with a service loading rate of 4 gpm/ft^2 and a headloss rate of 2 psi/ft. A value of 20 psi was assumed as the maximum allowable headloss across each vessel for these design conditions.

The design criteria assume a design flow rate of 2500 gpm for each of the water quality conditions and types of resins evaluated in the engineering analysis. The design run time was based on an estimated number of operational bed volumes, as defined in the mathematical model (discussed in Chapter 2) and presented in Table 3-1.

Table 3-1
Bed volumes used in engineering analysis

Resin Type	Low Nitrate	High Nitrate	Low Sulfate	High Sulfate
Type 1	9,770	6,900	36,500	2,600
TEA	81,600	26,300	113,400	22,300
TPA	111,700	28,500	120,800	30,600
Bifunctional	184,500	34,900	146,600	50,500

Capital Costs

Capital costs for a system are design and manufacturer specific and therefore critical to calculating the system cost. Included in this section are the cost per ion-exchange vessel (assumed at $231,000), auxiliary equipment costs including the MCC and PLC ($150,000), and pre-filter or auto-backwashing strainer ($60,000) to remove any suspended solids prior to the ion-exchange plant. It is assumed for the purpose of the cost estimate that the equipment has an expected life of 20 years. However, this will vary with capacity, water quality, maintenance, and use. The building square footage is the footprint calculated from the sum of all equipment requiring enclosure with an additional allowance for maintenance and general access. Once the facility footprint has been determined, the total square footage is multiplied by the assumed cost per square-foot ($150/ft^2), which was based on a building which includes HVAC and lighting.

Annual O&M Costs

Annual O&M costs account for the expense associated with operating and maintaining the ion-exchange system. The variables influencing the O&M costs include utilities (power at $0.12/kW-hr), daily maintenance and operation of equipment (assumed at 2 hours per day), equipment efficiency (70% for all pumps), annual resin replacement cost, and resin destruction costs ($2,500/20 ton).

Labor cost associated with operating an ion-exchange plant is estimated to be minimal due system automation. Resin replacement will require the most operator attention and therefore, the number of replacements per year will affect the labor cost. A labor rate of $85/hour was assumed for all plant staff.

The resin costs used to estimate the annual resin replacement costs are listed Table 3-2. These costs were used to estimate both the initial resin cost (included in the initial/startup capital costs) and annual replacement cost. These costs are list price only and do not include service agreements or contracts provided by some manufacturers for delivery of the resin, removal of the spent resin, and drying of the spent resin.

The resin cost presented in Table 3-2 show the ranges used in the engineering analysis and indicate an economy of scale based on the resin exhaustion rate or operable number of bed volumes per year. These costs represent the average list prices provided by a number of resin manufacturers, vendors, or distributors.

Table 3-2
Resin costs utilized in conceptual cost estimate

Type of Resin	Cost ($/ft^3) < 1000 ft^3	> 1000 ft^3
Type 1	$125	$92
TEA	$195	$165
TPA	$250	$200
Bifunctional	$1250	$750

Construction Related Costs

Construction related costs include engineering design, site work, legal, and administrative work involved in the construction of a new IX facility. These values are calculated using a range of construction related costs (as a percentage of the capital cost), based on experience with water treatment plant construction (Table 3-3).

Table 3-3
Range of construction related costs

Construction Related Cost	Average Range*	Values used in Engineering Analysis
Civil Site Work	1 to 10%	5%
Instrumentation	3 to 15%	8%
Electrical Site Work	7 to 12%	8%
Piping	5 to 12 %	5%
Construction Contingency	10 to 35%	20%
Contractor Overhead and Profit, Bonds, and Insurance	10 to 20%	10%
Engineering, Legal and Administrative	15 to 30%	8%

* Average ranges based on experience with surface water treatment plant design.

Construction related costs are largely site specific. For example, if the ion-exchange plant is located in a remote area, the range of construction costs may be higher than the data presented in Table 3-3. Transportation and labor rates may also be higher if labor and materials are not readily available in a remote location.

Construction contingencies are applied to the cost estimate to account for items not specifically included in a project scope but found to be necessary. The level of contingency selected should reflect the level of detail provided during pre-design. A low contingency budget reflects a high degree of confidence in the pre-design and a high contingency budget reflects a low level of confidence in the pre-design. A low degree of confidence may be due to limited availability of detailed costs or an experimental treatment technology. The recommended contingency levels for the varying types of cost estimates are listed in Table 3-4.

Table 3-4
Recommended contingency for corresponding level of estimate

Type of Cost Estimate	Level of Accuracy	Recommended Contingency
Order-of-Magnitude	+50% to –30%	20% to 30%
Conceptual	+40% to –20%	20% to 15%
Preliminary Design	+30% to –15%	15% to 10%
Definitive	+15% to –5%	10% to 5%

Assumptions Associated with Construction Costs

The following assumptions were made for the equipment and installation in the construction cost estimate:

- Installation includes integrity testing at each unit,
- Interconnecting pipe work complete with all necessary flexibility and anti-vibration equipment,
- Control system and interface to existing or other control and telemetry systems,
- Motor control center (MCC) and distribution boards to supply all items of electrical plant - electrical system to include cabling, trays, trunking, accessories and supports between the ion-exchange plant MCC, process control panel, motor drives and instrumentation,
- All necessary protective coatings and linings for plant,
- All ratings plates and labels,
- Feed and treated water quality instruments, and
- Feed water pumps.

Not included with the equipment and installation are:
- Final effluent feed pumps, and
- Feed water balance tank(s).

Present Worth Analysis

The present worth (present value) of the total capital costs to construct an ion-exchange plant, including the annual operation and maintenance needs, were evaluated by present worth analysis. The present worth of any value (such as the capital and O&M costs of the ion-exchange plant) is the equivalence of any future amount to the present amount. In this case it would be applied to the loan (or bond) estimate required to construct and operate the facility. The present worth is calculated by:

$$P = F(1 + i)^{-n} = \frac{F}{(1 + i)^{n}}$$

Where:
P = Present value
$(1+i)^{-n}$ = Single payment present worth factor
F = Future Amount
i = Interest rate
n = Number of years

Water Production Cost

The water production cost is the cost per unit (i.e., per gallon, thousand gallons, acre-ft, etc.) to produce water. Water production costs calculated in this engineering analysis assume an amortized capital cost plus the annual O&M cost. Understanding that preferences change between regions (including units) the costs are presented as both \$/1000 gallons and \$/acre-ft.

RESULTS OF ENGINEERING ANALYSIS

An engineering analysis was conducted to determine capital, operation, and maintenance costs associated with removing perchlorate from groundwater using a 2500 gpm single-use ion-exchange system. As discussed previously in Chapter 2, site-specific water quality (specifically nitrate and sulfate) will have a significant impact of the use and run length of single-use resins. To evaluate the impact of water quality on treatment costs, four different water quality conditions were evaluated (low nitrate, high nitrate, low sulfate, high sulfate). Perchlorate, bicarbonate, and chloride levels where held constant while the sulfate and nitrate levels were varied to determine the impact of these two variables on the water production cost. The specific water quality conditions used in the engineering analysis are presented in Table 3-5.

Table 3-5
Water quality conditions evaluated

Parameter	Low Nitrate	High Nitrate	Low Sulfate	High Sulfate
Perchlorate, μg/L	20	20	20	20
Nitrate, mg/L	1.9	87	19	19
Sulfate, mg/L	48	48	4.8	192
Chloride, mg/L	21	21	21	21
Bicarbonate, mg/L	122	122	122	122

Using the results of the mathematical modeling efforts (Chapter 2), the number of bed volumes of operational run time allowed before perchlorate breakthrough occurred were determined for each water quality condition and ion-exchange resin type (Type 1, TEA, TPA, and bifunctional). The number of bed volumes used in the engineering analysis was previously presented in Table 3-1.

Summary of Treatment Costs

The total water production cost (\$/1000 gal and \$/acre-foot) for each alternative are presented in Figure 3-1 and include the annual O&M cost and the amortized capital costs. Table 3-6 provides a summary of the estimated O&M while the capital costs are provided in Table 3-7. The O&M costs are affected by a number of variables including (1) the water quality which affects the operational run time or number of bed volumes, and (2) the type of resin used which affects replacement cost and incineration cost. Site-specific conditions and O&M factors may dictate a different design for each water treatment site, but for the purpose of this engineering analysis the water treatment plant design is assumed to be the same for each system.

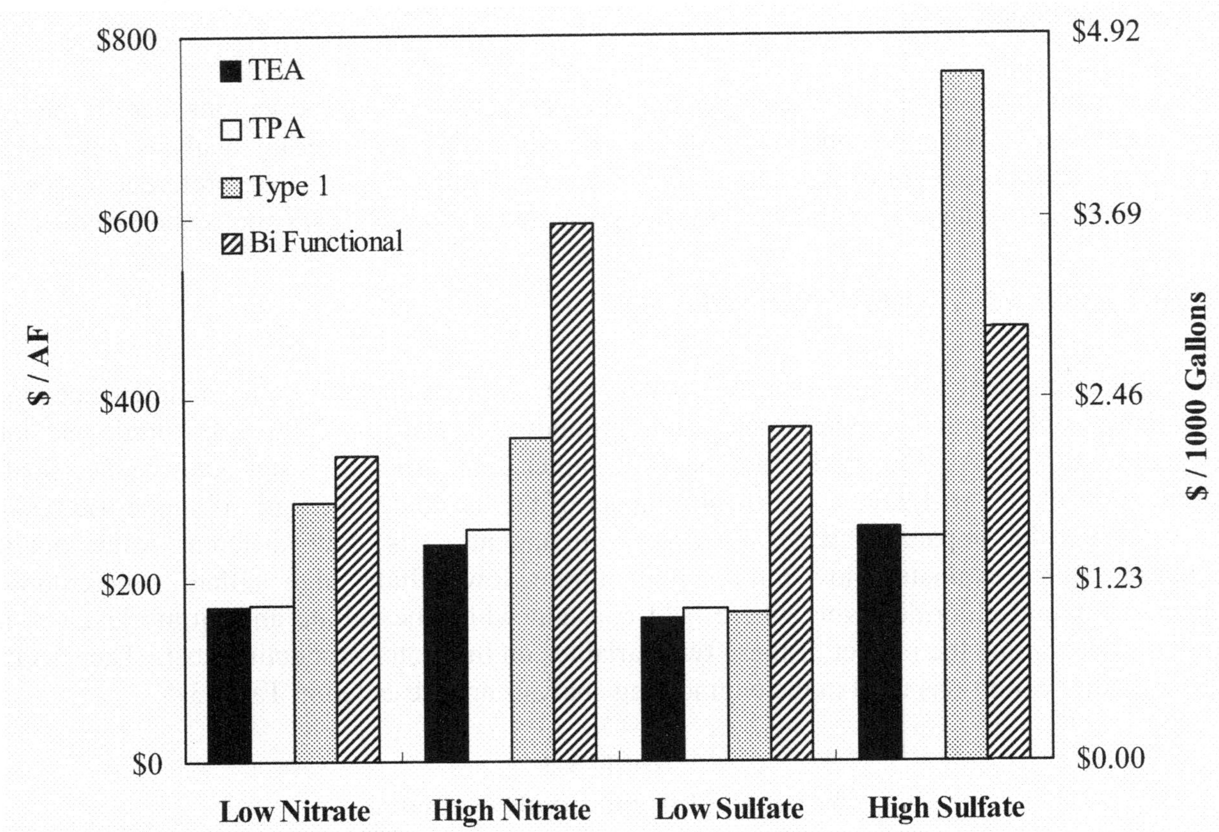

Figure 3-1 Effect of variable water quality on water production cost

From an evaluation of Figure 3-1 and the data presented in Table 3-6, the greatest impact on water production cost is the resin replacement cost. The resin replacement cost is affected by the water quality (number of operated bed volumes) and the resin type selected for a given water quality condition. For example, the high nitrate and high sulfate conditions were more favorable (cost-wise) to the TEA and the TPA resins. However, the costs were still high relative to the low nitrate and low sulfate water quality conditions for all of the resin types. Therefore, systems with high levels of either nitrate or sulfate might dictate a different resin all together from the four types of resins evaluated in this study (based on the water production cost).

Additionally, comparing the data in Table 3-1 (bed volumes used in engineering analysis) to Figure 3-1, the system with the longest run time (highest number of operable bed volumes) may not be the most cost effective, particularly if the resin unit cost ($\$/ft^3$) offsets the advantages of a longer run time prior to exhaustion.

Table 3-6
Estimated O&M costs per resin type*

O&M Parameter	Type I Low Nitrate (9,770 BV)†	Type I High Nitrate (6,900 BV)†	Type I Low Sulfate (36,500 BV)†	Type I High Sulfate (2,600 BV)†
Annual Resin Costs	$660,000	$930,000	$180,000	$2,500,000
Energy	$16,000	$16,000	$16,000	$16,000
Labor	$62,000	$62,000	$63,000	$62,000
Annual Incineration Cost	$22,000	$31,000	$5,800	$82,000
Total Cost per Year	$760,000	$1,000,000	$260,000	$2,700,000
Cost per Acre-Foot	$190	$250	$65	$670
Cost per 1000 gallons	$0.60	$0.80	$0.20	$2.10

O&M Parameter	TEA Low Nitrate (81,600 BV)†	TEA High Nitrate (26,300 BV)†	TEA Low Sulfate (113,400 BV)†	TEA High Sulfate (22,300 BV)†
Annual Resin Costs	$170,000	$440,000	$120,000	$520,000
Energy	$16,000	$16,000	$16,000	$16,000
Labor	$64,000	$63,000	$65,000	$63,000
Annual Incineration Cost	$2,600	$8,100	$1,900	$9,500
Total Cost per Year	$250,000	$530,000	$200,000	$600,000
Cost per Acre-Foot	$60	$130	$50	$150
Cost per 1000 gallons	$0.20	$0.40	$0.15	$0.45

O&M Parameter	TPA Low Nitrate (111,700 BV)†	TPA High Nitrate (28,500 BV)†	TPA Low Sulfate (120,800 BV)†	TPA High Sulfate (30,600 BV)†
Annual Resin Costs	$160,000	$490,000	$140,000	$460,000
Energy	$15,700	$15,700	$15,700	$15,700
Labor	$65,000	$63,000	$65,000	$63,000
Annual Incineration Cost	$1,900	$7,500	$1,800	$6,900
Total Cost per Year	$240,000	$580,000	$220,000	$550,000
Cost per Acre-Foot	$60	$140	$50	$140
Cost per 1000 gallons	$0.20	$0.45	$0.15	$0.45

O&M Parameter	Bifunctional Low Nitrate (184,500 BV)†	Bifunctional High Nitrate (34,900 BV)†	Bifunctional Low Sulfate (146,600 BV)†	Bifunctional High Sulfate (50,500 BV)†
Annual Resin Costs	$470,000	$1,500,000	$600,000	$1,000,000
Energy	$15,700	$15,700	$15,700	$16,000
Labor	$67,000	$63,000	$65,000	$63,000
Annual Incineration Cost	$1,100	$5,700	$1,400	$3,900
Total Cost per Year	$550,000	$1,600,000	$680,000	$1,100,000
Cost per Acre-Foot	$140	$400	$170	$270
Cost per 1000 gallons	$0.45	$1.20	$0.50	$0.85

*Cost rounded to two significant figures.

† Represents number of bed volumes to perchlorate breakthrough

Table 3-7
Estimated capital costs*

Capital Parameter	Estimated Cost
Vessel Cost	$924,000
Vessel Related Support & Piping	$462,000
Vessel Installation Costs	$693,000
Pump Costs	$40,000
Building Costs	$330,000
Sub- total 1 - - Equipment Total	$2,889,000
Sub-total 2 - - Construction Cost	$1,040,000
Sub-total 3 - - Construction Cost + Equip. Cost	$3,929,000
Engineering, Legal, & Admin	$314,000
Contingency	$393,000
Cost to Construct Project	$4,636,000
Cost per Acre-Foot	$90
Cost per 1000 gallons	$0.30

*The capital costs that are presented do not include the initial resin cost. However, the initial resin costs are included in the water production cost presented in Figure 3-1.

SINGLE-USE RESIN VERSUS CONVENTIONAL ION-EXCHANGE TREATMENT

A selected water quality condition was used to compare the costs of a single-use IX system to a conventional regenerable IX process with and without brine treatment and reuse (Figure 3-2). The conventional IX system with brine treatment assumes biological treatment and reuse of spent brine while the conventional IX system without brine treatment assumes the waste brine is sent to a dedicated brine line sewer.

As with the single-use system, the conventional IX costs were calculated from theoretical systems. Both the single-use and conventional regenerable ion-exchange costs were based on the following water quality conditions to be representative of a typical groundwater in southern California: perchlorate (20 µg/L) nitrate (25 mg/L), sulfate (48 mg/L), chloride (21 mg/L), and bicarbonate (122 mg/L). The conventional IX plant (both with and without brine treatment and reuse) assumed that a regenerable polyacrylic resin was used.

Comparing the cost of using a single-use resin facility to a conventional IX facility (either with or without brine treatment and reuse) indicated that the water production cost is still both water quality and resin cost specific. For the given water quality condition, a disposable resin plant using Type I at $0.95/1000 gallons ($300/AF), TEA $0.55/1000 gallons ($180/AF), or TPA $0.60/1000 gallons ($190/AF) were estimated to be more cost-effective than a conventional IX system at $1.25/1000 gallons ($400/AF) or conventional IX with brine treatment at $1.00/1000 gallons ($325/AF). The bifunctional resin approach was found to be a more costly alternative at $1.45/1000 gallons ($475/AF), although the resin would allow for longer operational run length or bed volumes. The major cost contribution for a bifunctional resin system was the cost of replacement resin.

It is important to note that the data presented in Figure 3-2 was based on a water quality with moderate nitrate (25 mg/L) and sulfate (49 mg/L) levels. In comparison, data presented in Figure 3-1 included some cases with high levels of nitrate (87 mg/L) or sulfate (192 mg/L).

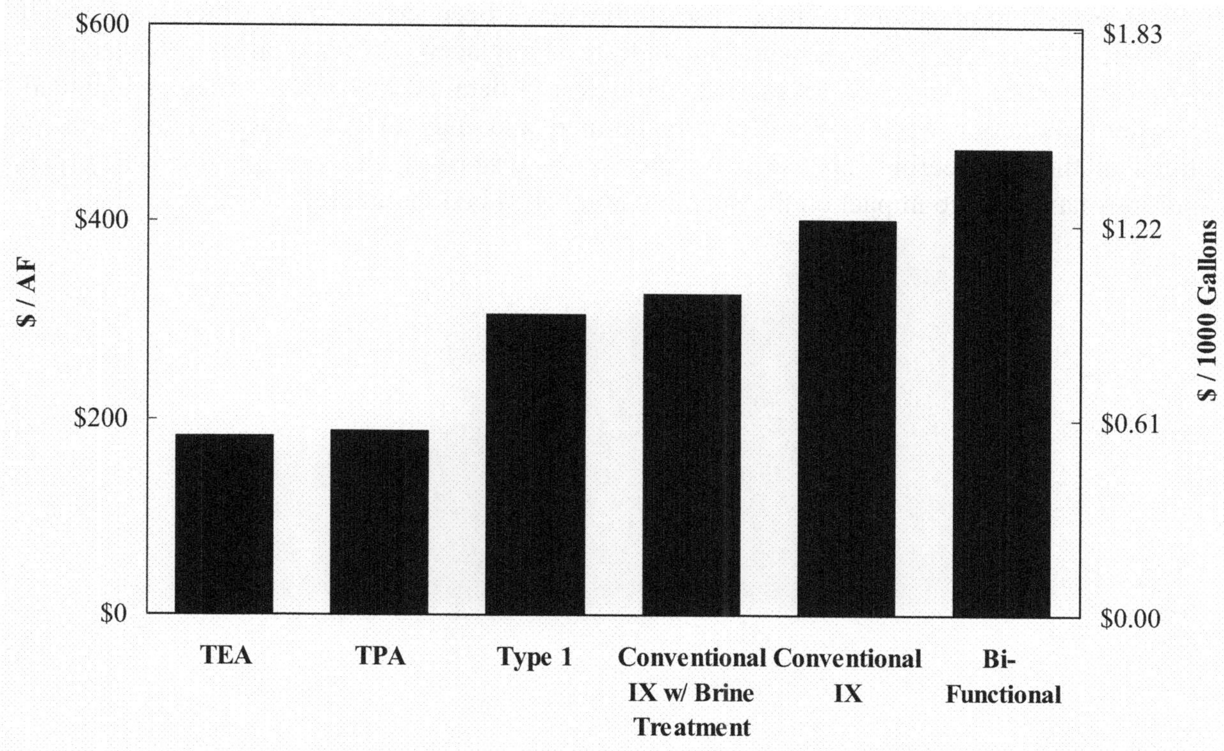

Figure 3-2 Disposable resin versus conventional ion exchange

The O&M costs for a disposal type IX system, regardless of the resin or water quality selected constitutes a considerable portion of the water production costs. The annual resin replacement is the greatest portion of the annual O&M costs, and is largely effected by the water quality but is also effected by the resin cost per cubic foot.

Assuming the capital costs are the same for each type of condition evaluated, selecting the most cost effective IX system is dependent upon matching the water quality conditions with the appropriate type of application. In other words, one must consider all the options including disposable resin IX system versus conventional IX system (with or without brine treatment), and the type of resin for a given water quality condition.

For example, under high nitrate conditions, an alternative resin (other than those presented in this study) would be selected that is nitrate selective or a regenerable system would be appropriate. The cost of using a bifunctional resin under high nitrate conditions is $1.70/1000 gallons ($550/AF) is far less economical compared to the TEA at $0.75/1000 gallons ($240/AF). Therefore, matching the water quality to the right resin is key for optimizing the water production cost. Additionally, under the one water quality condition evaluated in this study for comparing disposable resin applications to conventional IX (either with our without brine treatment), the bifunctional resin was the only disposable resin application that was more costly than the conventional IX systems.

Two case studies have been provided in the following chapter to give “real-world” examples of disposable resin applications. These case studies have been provided by vendors who supply full-service agreement contracts which include resin replacement, removal of exhausted resin, drying of the resin, and resin destruction. However, a detailed list of the variables included in the water production costs are not clearly defined and may or may not include parameters defined in the engineering analysis. As previously discussed, the site-specific water quality conditions have a large impact on the operable number of bed volumes.

CHAPTER 4 – CASE STUDIES

Within the past two years, several public sites in California have utilized single-use resins for perchlorate treatment. The two primary system suppliers presently involved with full-scale single-use resin treatment system are US Filter and Calgon Carbon Corporation (Calgon). Typically, the system supplier provides a complete turnkey contract to the utility, which includes ion-exchange pressure vessels, perchlorate-selective resin, and resin destruction through incineration or co-generation waste-to-energy disposal. This chapter summarizes information solicited from US Filter and Calgon with respect to water quality, total treatment costs of full-scale systems, and historical experience with perchlorate treatment using ion exchange. The "real-world" findings presented below were used to validate the conceptual-level engineering analysis presented in Chapter 3.

CALGON CARBON CORPORATION

Responding to a market need for a simple, operator and maintenance-free perchlorate treatment system, Calgon Carbon Corporation has been actively developing and optimizing the use of fixed-bed, highly perchlorate-selective ion-exchange resins for the removal of perchlorate from contaminated groundwater. The initial application of the fixed-bed single-use systems was geared towards applications where a short-term low capital cost solution was required in order to evaluate and implement a long-term treatment solution. The very first application was applied to a groundwater remediation project where the perchlorate levels of 30-200 ppm were orders of magnitude higher than the levels typically seen in most of the drinking water applications in Southern California. The second application was for California Domestic Water Company (Whittier, CA) in which the perchlorate levels were expected to be in the 10-20 ppb range. California Domestic, the first application of the fixed-bed single-use approach for a drinking water, was commissioned in July 2002. California Domestic chose to pursue the fixed-bed single-use option even though the operating costs were significantly higher than a regenerable ion-exchange system (<$525/acre foot guarantee vs. $160/acre foot) because there was no brine line available to dispose of the waste brine. California Domestic was also not sure how long a perchlorate treatment system would be required as other systems scheduled to start up were expected to reduce perchlorate levels below the action limit.

Since the initial application at California Domestic, the fixed-bed single-use approach appears to be the solution of choice for treating perchlorate-contaminated groundwaters for drinking water applications. This trend resulted from the development and use of resins with higher perchlorate selectivity with significantly lowered operating costs compared to the regenerable ion-exchange approach.

Table 4-1 summarizes Calgon's major perchlorate treatment installations to date in California. There are a total of seven single-use plants currently in operation, with at least three additional plants in construction and expected to be operational by the end of 2004.

Table 4-1
List of Calgon Carbon Corporation's major perchlorate treatment installations

Client Name	Location	Flowrate (gpm)	Startup Date
Operating			
California Domestic Water Company	Whittier, CA	5,000	July 2002
City of Riverside	Riverside, CA	5,000	December 2002
City of Riverside	Riverside, CA	2,000	May 2003
West San Bernardino County Water District	Rialto, CA	2,000	May 2003
Confidential	CA	1,000	June 2002
City of Monterey Park	Monterey Park, CA	4,050	July 2003
Confidential	CA	2,000	October 2003

California Domestic is very typical of the water quality encountered in perchlorate applications located in the Southern California region. A comparison of the water quality at California Domestic relative to several other locations in the area is provided in Table 4-2. Using the data provided in Table 4-2, comparisons of capital and operating costs were determined for four different technology options including fixed-bed single-use using a standard Type I resin, a perchlorate-selective resin, a regenerable ion-exchange system (Calgon ISEP), and a regenerable ion-exchange system with brine treatment and reuse (Calgon ISEP+). The capital and operating costs for each of these four options are shown in Table 4-3. The first fixed-bed single-use system installed at Cal Domestic used a standard Type I resin. The operating cost guaranteed for this system was <$525/acre-foot and the system is currently operating at $260/acre-foot.

Table 4-2
Typical water qualities of Calgon single-use systems in southern California

Client	Flow Rate (gpm)	ClO_4^- (ppb)	SO_4^{2-} (ppm)	NO_3^- (ppm)	Cl^- (ppm)	CO_3^- (ppm)
California Domestic	5000	<20	40	30-40	20-40	250
City of Riverside	5000	<20	45-60	10-25	40	230
West San Bernardino	2000	<20	30	30	10	190
City of Monterey Park	4050	<20	45	5-20	20-40	300

Table 4-3
Comparison of capital and operating costs for different resin based technology options using California Domestic water quality

	Fixed-bed single-use (standard Type I resin)	Fixed-bed single-use (perchlorate-selective resin)	ISEP	ISEP+
Annual Operating Cost	$260/AF	$175/AF	$140/AF	$130/AF
Capital Cost	$1.4 M	$1.4 M	$4.4 M	$9.9 M
Water Production Cost†	$289/AF	$203/AF	$227/AF	$326/AF
	$0.89/1000 gal	$0.62/1000 gal	$0.70/1000 gal	$1.00/1000 gal

† Amortized for 20 years at 5 percent interest

As shown in the Table 4-3, using a perchlorate-selective resin would further reduce the operating cost of the system to $175/acre-foot. While this is 20 to 25 percent higher than the ISEP and ISEP+ options, the capital cost of the system is 1/3 to 1/5 of the ISEP and ISEP+ costs. In addition to the capital cost savings, the lead-time from purchase to start-up of the fixed-bed single-use systems is significantly shorter (6 months) than the regenerable options (12 months). As a result, it is easy to see why the trend in the market moving towards the fixed-bed single-use systems. The limitation of the fixed-bed single-use system, however, is that it cannot be used where nitrate removal is required in addition to perchlorate removal. In this case, a regenerable ion-exchange system would typically be used (i.e., ISEP). If a brine line is not available, then the ISEP+ would be the technology of choice.

US FILTER

US Filter also offers turnkey solutions, ranging from design and installation of treatment systems through system operation and disposal of media. Unlike Calgon Carbon Corporation, US Filter does not currently have any conventional full-scale ion-exchange plants in operation using the regenerable resin approach. Instead, US Filter has been actively promoting the use of single-use resins for perchlorate removal and resin destruction.

Table 4-4summarizes the known US Filter perchlorate treatment installations in California. Single-use system installations to date include sites that provide potable drinking water from municipal supply wells as well as pump and treat remediation sites, and typically utilize Type I styrenic resins. Plant capacities range from very small (100 gpm) to larger systems (>2000 gpm) and perchlorate treatment can be achieved through a service contract which include either temporary portable exchange vessels or fixed on-site vessels.

Table 4-4
List of USFilter major perchlorate treatment installations

Client Name	Location	Flowrate (gpm)	Startup Date
Operating			
City of Colton	Colton, CA	4,000	October 2003
Fontana Water Company	Fontana, CA	5,000	January 2004
City of Morgan Hill (SCVWW)	Morgan Hill, CA	1,500	July 2003
City of Rialto	Rialto, CA	1,500	September 2003
West San Martin Water Works	San Martin, CA	1,250	August 2003
West San Martin Water Works	San Martin, CA	2,000	November 2003
West Valley Water Company	San Bernardino, C.	2,000	June 2003
Aerospace Company	Sacramento, CA	1,000	May 2003
Aerospace Company	Sacramento, CA	1,500	December 2003
Flare Mfg. - West San Martin	San Martin, CA	200	June 2003

As shown in Table 4-4, US Filter is currently operating single-use systems throughout California. Table 4-5 presents water quality data for selected US Filter single-use installations. Sites that focus on groundwater remediation, such as the Aerospace company in the Sacramento area, treat much higher concentrations of perchlorate (up to 270 ppb) than that which is typically observed in southern California groundwaters (~20 ppb). As previously discussed (Modeling Results section), the concentration of perchlorate does not have a significant impact on the operation run length of single-use system and as a result the operating costs of these systems are very similar.

Table 4-5
Typical water qualities and annual treatment costs of US Filter single-use systems in California

Client	Flow Rate (gpm)	ClO_4^- (ppb)	SO_4^- (ppm)	NO_3^- (ppm)	Annual Operating Cost
City of Morgan Hill	1500	<20	44	24	$246/AF
West San Bernardino	2000	<20	35	26	$230/AF
Aerospace Company	1000	270	10	20	$240/AF

†Amortized for 20 years at 5 percent interest

SUMMARY

Perchlorate treatment options have rapidly evolved since the late 1990's when the problem was initially discovered. The initial approach focused on regenerable ion-exchange systems. The fixed-bed single-use approach was entertained initially as only a short term or temporary solution. Over a period of a few short years, significant research has been invested in developing much more selective resins that have lead to significant cost reductions in the single-use operating costs. The current operating cost of single-use resins average \$200 - \$300/AF based on existing full-scale plants with varying water qualities. As single-use systems are quickly becoming a cost-effective alternative treatment option, both Calgon Carbon and US Filter are exploring new ways to drive perchlorate treatment costs down through continued development of the perchlorate treatment technology.

CHAPTER 5 – CONCLUSIONS

The detection of perchlorate in California groundwaters in early 1997, prompted studies to evaluate a wide variety of perchlorate treatment technologies. The technologies demonstrated to be technically feasible for drinking-water treatment applications are biological reduction, ion exchange, reverse osmosis (RO) membranes, and pretreated granular activated carbon (GAC). Of these technologies, ion exchange has been the most proven and publicly accepted process capable of meeting existing treatment requirements.

The continued exploration of ion-exchange perchlorate treatment has resulted in an increasing interest and implementation of single-use or "non-regenerable" ion-exchange resin systems. These systems generally employ resins with a high affinity for perchlorate in a simple fixed-bed configuration. Once exhausted, the resin is simply removed and replaced with fresh resin. The exhausted resin is then either regenerated at a centralized processing facility that is designed to manage the challenging regeneration procedure with potentially hazardous chemicals, or is destroyed in an incineration or co-generation facility.

This report has evaluated the four types of resins most commonly considered for perchlorate removal and their applicability towards single-use systems. These resins, generally classified as strong base anion (SBA) resins, included: Type I, triethyl, tripropyl, and bifunctional resins that contain two different quaternary amine groups in their structure. Based on modeled performance of these four resins on a range of water quality conditions, an engineering analysis was performed to estimate the water production costs associated with these systems. This analysis was followed by a case study describing the full-scale implementation of a single-use system and the costs of this system compared to regenerable ion-exchange systems. The engineering analysis was followed by two real-world case studies of disposable resin applications for perchlorate removal. These case studies included water production costs for each site.

This is a rapidly developing field. New products are coming to market, prices for products are continuously changing, and new approaches are being developed based on these new experiences in the field. Consequently, the results presented in this report reflect the information, prices, resins, and operational conditions present at the time the report was prepared and reflect only those conditions. It is urged that each individual situation be evaluated in a manner similar to that taken in this report so that an accurate assessment of single-use resin system use can be established.

Based on this knowledge contained within this report, the following conclusions can be made about single-use ion-exchange systems for perchlorate treatment:

- Relative impact of selected water quality parameters on effective life of resins. The volume of water or duration of time that a single-use resin system can effectively remove perchlorate is relatively less sensitive to the perchlorate concentration (1 to 200 μg/L) as compared to concentrations of chloride (5 to 200 mg/L), nitrate (0.5 to 20 mg/L as N), or sulfate (5 to 200 mg/L).

- Relative performance of resins in model waters. For all four model water qualities (see Table 3-5) evaluated, it was determined that the bifunctional resins consistently had a longer effective life than the triethyl and tripropyl resins, which, in turn, had longer lives than the Type I (trimethyl) resins. It was estimated that tripropyl resins had a 37 percent longer effective life than triethyl resins in low-nitrate or high-sulfate conditions, whereas they only lasted about 8 percent longer in high-nitrate or low-sulfate conditions. Furthermore, the triethyl resins lasted 3 to 9 times longer than the trimethyl resins for equivalent water qualities.

- Relative water production costs of single-use systems. The engineering analysis of each resin under the range of water qualities demonstrated that the resin with the longest effective life was not necessarily the most economical (see Figure 3-1). For three of the four model waters, the expensive bifunctional resin had the highest water production costs. The triethyl and tripropyl resins generally had the lowest costs. However, for low-sulfate conditions, the Type I (trimethyl) resin was cost competitive.

- Relative costs of single-use versus conventional ion-exchange systems. For a water quality typical of those encountered in southern California (ClO_4^- <20 μg/L, SO_4^{2-} 40 mg/L, NO_3^- 9 mg/L as N, Cl^- 30 mg/L, HCO_3^- 250 mg/L), it was estimated that single-use systems are typically more economical than regenerative systems (see Table 4-3).

- Single-use ion-exchange systems are being utilized. Within California there are at least 17 full-scale installations utilizing single-use ion-exchange systems with a total design capacity of nearly 60-MGD.

With the increasing demands on water supplies, the increasing detection of perchlorate-impaired waters, and the development of perchlorate regulations in several states, single-use ion-exchange resins have been shown to be more than just an effective interim measure. Depending on water quality the costs of these single-uses can be substantially lower than conventional ion-exchange systems for perchlorate removal.

REFERENCES

Aldridge, L., G. Lehman, T. Gillogly, D. Clifford, D. Roberts, X. Lin. 2004. *Treatability of Perchlorate in Groundwater using Ion Exchange Technology*. AwwaRF, Denver, CO.

Clifford, D., D. Roberts, A. Tripp, Y. Cang, L. Aldridge, T. Gillogly and L. Boulos 2003. *Treatment of Perchlorate in Groundwater by Ion Exchange Technology*. AwwaRF, Denver, CO.

Dorfner, L. 1991. Synthetic Ion Exchange Resins. In *Ion Exhangers*. Edited by K. Dorfner. Berlin: Walter de Gruyter.

Harland, C. E. 1994. Ion Exchange: Theory and Practice. Cambridge, United Kingdom: The Royal Society of Chemistry.

Helfferich, F. 1962. Ion Exchange. New York: McGraw Hill

Li, P., and A.K. SenGupta. 2000. Intraparticle Diffusion During Selective Ion Exchange with a Macroporous Exchanger. *React. Polym.,* 44:273-287.

Subramonian, S., D.A. Clifford. 1988. Monovalent/divalent selectivity and the charge separation concept. *Reactive Polymers*. 9:195-209.

Tripp, A.R., D.A. Clifford. 2000. “The Treatability of Perchlorate in Groundwater using Ion-Exchange Technology.” In *Perchlorate in the Environment*. Edited by E.T. Urbansky. New York: Kluwer Academic/Plenum Publishers.

ABBREVIATIONS

Awwa	American Water Works Association
AwwaRF	Awwa Research Foundation
BV	Bed volume
BTU	British thermal unit
°C	Degrees celsius
Cl^-	Chloride
ClO_4^-	Perchlorate
EBCT	Empty bed contact time
ft	Feet
GAC	Granular activated carbon
gpm	Gallons per minute
IX	Ion Exchange
L	Liter
lb	U.S. pound
M	Molar
MCC	Motor control center
MCL	Maximum Contaminant Level
mg/L	Milligram per liter
min	Minute
mL	Milliliter
MWH	Montgomery Watson Harza
µg/L	Microgram per liter
NaCl	Sodium chloride
O&M	Operation and maintenance
PHG	Public health goal
PLC	Programmable logic controller
RO	Reverse osmosis
SBA	Strong base anion

IWA Publishing's authorised EU representative for General Product Safety Regulations is Diane D'Arras, 15 rue Duret, 75116 Paris, France, e-mail: safety@iwap.co.uk.

Printed and bound by CPI Group (UK) Ltd, Croydon, CR0 4YY
17/09/2026
02218804-0009